Peter Stoll

Gynecological Vital Cytology

Function · Microbiology · Neoplasia
Atlas of Phase-Contrast Microscopy

With 145 Figures

Springer-Verlag
Berlin Heidelberg GmbH 1969

Author:

Prof. PETER STOLL, M.D., FIAC, Professor of Obstetrics and Gynecology of the University of Heidelberg, Director of the Department of Obstetrics and Gynecology, Mannheim Municipal Hospital

Collaborators:

GISELA DALLENBACH-HELLWEG, M.D., Assistant Professor of Pathology of the University of Heidelberg, Head of the Morphological Division, Department of Obstetrics and Gynecology, Mannheim Municipal Hospital

HEINZ GUNDLACH, Ph.D., member of the scientific staff of Carl Zeiss, Oberkochen

JOST JAEGER, M.D., Assistant Professor of Obstetrics and Gynecology of the University of Heidelberg, Assistant Director of the Department of Obstetrics and Gynecology, Mannheim Municipal Hospital

English translation:

FREDERICK DALLENBACH, M.D., Associate Professor of Pathology, Department of Experimental Pathology, German Cancer Research Center, Heidelberg

Spanish translation:

RAUL LARRAGUIBEL, M.D., Department of Obstetrics and Gynecology, University of Santiago de Chile

Picture on book jacket:
"near an epithelial cell a trichomonad
with two flagella in front and a
trailing flagellum. Magnification about $\times$ 1,000".

ISBN 978-3-662-23580-5   ISBN 978-3-662-25659-6 (eBook)
DOI 10.1007/978-3-662-25659-6

Library of Congress Card Number 74-99013.

Title-No. 1637

Preface

In gynecological practice, techniques of examination are being supplemented more and more by cytodiagnosis. Thus it has become necessary
to acquaint the gynecologist of the possibilities, use and limits of cytodiagnosis. Such is the purpose of the book "Gynäkologische Cytologie"
(Stoll, Jaeger, Dallenbach, Springer-Verlag 1968).
In general, the practicing gynecologist will merely make the vaginal,
ecto- and endocervical smears and leave the diagnosis of them to a cytological laboratory. Only in rare cases will a trained and experienced
specialist set up his own cytological laboratory for outpatients, although
such undertaking would be very desirable for propagating the cytological method.
The cytological analysis of unstained fresh smears during the gynecological examination allows an immediate study to be made of microflora and cellular atypia. For such cytological studies microscopes are
employed in which a high-contrast image of the specimen is obtained
by optical means (phase-contrast and interference-contrast microscopy),
thereby eliminating the need for fixation and staining.
In the nineteen-thirties the Dutch physicist Zernike investigated the
formation of high-contrast images of transparent objects by modifying
the path of light. In 1941, his ideas were put into practice by the firm of
Carl Zeiss, Jena. Zernike received the Nobel Prize for physics in 1953.
The method he had developed proved of great value in biology and in
medicine, above all for the examination of living objects. It was introduced into gynecology by Runge, Vöge, Haselmann and Zinser in 1949.
Since then a wealth of experience has been gathered with the method,
so that its usefulness for rapid diagnosis in the consulting room has been
established beyond any doubt. A prerequisite, of course, is the doctor
must become familiar with cytology and practice the method continously. The same applies to the Nomarski interference contrast. In addition to preparing the usual smears by the Papanicolaou technique, we
use the phase-contrast method routinely at our hospital. Over the last
thirteen years the limits of the method have been explored and photographic records of over 1,000 specimens collected. Our photomicrographs
were produced in the Heidelberg and Mannheim gynecological hospitals

(technical supervision: Otto Krieger). Discussions with former collaborators, among them Prof. Dr. H. Bach of Pforzheim, Primarius Dr. O. Ledermair of Linz, Dr. D. Francke of Bruchsal, Dr. H. Bachmeyer of Karlsruhe, Dr. O. Brunner of Salzburg, Dr. I. Delnon of Bern and Dr. M. Yilmaztürk of Istanbul, confirmed our opinion that the phase-contrast method is practical for general clinical use. Therefore in the textbook mentioned above, in addition to conventional cytology according to Papanicolaou, the phase-contrast method is explained in detail and presented here in the form of an atlas. Prof. Dr. W. Wundt, Director of the Department of Microbiology of the Mannheim Municipal Hospital, was of great assistance in bacteriological questions. For the English translation we owe thanks to Prof. Dr. F. Dallenbach of the German Cancer Research Center, Heidelberg; for the Spanish translation to Dr. R. Larraguibel of the Gynecological Hospital of the University of Santiago, Chile.

The authors trust that this atlas will not only appeal to cytologists interested in scientific questions, but that it will above all find its way into the gynecological consulting room. May it aid the gynecologist in his work, deepen his understanding of cytodiagnostics, and assist him in his decisions on treatment.

P. STOLL
G. DALLENBACH-HELLWEG
H. GUNDLACH
J. JAEGER

Mannheim, November, 1969

Table of Contents

Part I. Practical Application 1
   Laboratory Cytology (Papanicolaou Test) 1
   Cytology at the Doctor's Office (Phase Contrast Microscopy). 1
A. Preparing the Specimen for Study 3
B. Use of Cytology during Clinical Examination 3
C. Evaluation of the Wet-Mount Preparation under the Phase-
   Contrast Microscope 4
     I. Diagnosis of Ovarian Function 4
      Phases of the Menstrual Cycle. 4
      The Pre-Ovulatory Smear 5
      The Post-Ovulatory Smear 5
      Pregnancy 5
      Atrophy 5
      The Androgenic Smear 5
      The Estrogenic Smear 5
     II. Diagnosis of Microorganisms 6
     III. Search for Tumor Cells 6
     IV. Single Cells. 7
      Characteristics of Single Cells in Wet-Mount Preparations 7
      Normal Cells 7
      Dyskaryotic Cells 8
      Carcinoma Cells 8
      Columnar Epithelium 9
     V. Secondary Changes in Cells 9
     VI. Artifacts 9
     VII. The Inadequate Smear 10
Part II. Technical Explanations 11
I. Modification of the Object 11
II. Modification of the Microscope Light Path 11
   Phase-Contrast Method 11
   Nomarski Differential Interference Contrast 12
   Equipment 13
   Microscopes for Gynecological Cyto-Diagnostics 13
   Photomicrographic Data 14
   Technical Illustrations 15

Part III. Cytological Illustrations 21
The Menstrual Cycle 22
Pregnancy 30
Smear Showing Androgen Effect 32
Atrophy 34
Hormonal Effect 38
Vaginal Flora — Döderlein's Bacillus 42
Microorganisms of the Vagina, Trichomonas 44
Microorganisms of the Vagina, Bacteria 48
Vaginal Flora — Mycoses 50
Vaginal Flora — Hemophilus Vaginalis 54
Vaginal Flora — Cocci 56
Vaginal Flora during Pregnancy 58
Endocervical Canal — Columnar Epithelium 60
Endocervical Canal — Columnar Epithelium Secondary Changes . 63
Ectropion — Squamous (Cell) Metaplasia 66
Dysplasia 68
Carcinoma 70
Sperm in Cervical Secretions 72
Cervical Mucus (the Fern-Leaf Test) 73
Artifacts 74

Pertinent references may be found in the textbook „Gynäkologische Cytologie"
(STOLL, JAEGER, DALLENBACH-HELLWEG, Springer-Verlag 1968).

I. Practical Application

Cytodiagnosis (the Papanicolaou test) has become a most important procedure in the gynecological examination. Study of the vaginal secretions is especially useful for:

a) evaluating ovarian function:
   By ascertaining the proliferative state of the vaginal epithelium, the state of hormonal function may be determined.
b) making a microbiological diagnosis:
   One may determine what microorganisms are in the vagina (bacterial, fungal, protozoan).
c) detecting malignant tumors, especially carcinomas:
   By searching for tumor cells in vaginal secretions, the preclinical carcinomas or precancerous dysplasias of the portio vaginalis may be detected early.

Laboratory Cytology (Papanicolaou Test)

Generally, during examination at the doctor's office, vaginal secretions are obtained either by suction, cotton applicators, or the Ayre spatula; and a smear is prepared which is then fixed and sent to the cytologic laboratory. Here the smear is stained, examined, and interpreted by a trained cytologist. The referring physician receives a report of the findings relevant to the three points listed above (a—c). With such a procedure the taking of the smear rests with the clinician, and the cytological diagnosis rests with the laboratory doctor; that is, in different hands. The procedure however takes time; the results of the cytological study are not immediately available to the clinician. Consequently, the patient must often return for a repeat examination.

Cytology at the Doctor's Office (Phase-Contrast Microscopy)

The clinician who has a basic knowledge of cytology should be able to evaluate a smear during the patient's visit. The phase contrast microscope has proved its value for "the rapid diagnosis during office hours". With the use of this microscope a quick inspection of a vaginal smear may be made while the patient is still on the examining table. The technique of phase contrast microscopy is especially suited for:

1. evaluating ovarian function from the vaginal epithelial cells, according to the morphological criteria set down by Papanicolaou.
2. obtaining an impression of the different microorganisms in the vagina. It is especially helpful for making a differential diagnosis of vaginal discharge and for controlling the success or failure of therapy.

The following organisms may be readily differentiated:

a) The Döderlein's bacillus.
b) Trichomonads.
c) Fungal infections (vaginal mycosis, especially Candida albicans).
d) Hemophilus vaginalis.
e) Mixed flora.

3. evaluating polymorphonuclear leukocytes and erythrocytes in vaginal and cervical secretions, when these are taken separately, with the aim of localizing the inflammation (vaginitis, endo-cervicitis, endometritis) and a source of bleeding (hemorrhagic colpitis, occult bleeding from the endocervical canal.

4. searching for tumor cells in the vaginal or cervical secretions (search for incipient cancer) according to the criteria for atypical and dysplastic cells as formulated by Papanicolaou.

5. determining the phase of the menstrual cycle by testing for the crystallization phenomenon of cervical mucus (Fern-Leaf test).

6. studying the mobility of sperm in the cervical secretions (Sims-Huhner test).

In addition, the method may be used for:

7. evaluating corpuscular constituents in urine and urinary sediment.

8. studying the fluid obtained from an ascites or from ovarian cysts.

The rapid method serves ideally as a guide, and is especially worthwhile for repeated check-ups. It has the advantage that the results of the cytological study are available while the patient is present; the results supplement the clinical and colposcopic studies and may be used directly for deciding treatment. The state of ovarian function and the types of microorganisms diagnosed may be evaluated with the clinical history and discussed with the patient. In each instance the likelihood is increased that proper therapy will be selected.

The search for carcinoma cells is possible since they, as in the fixed and stained Papanicolaou preparations, may be distinguished from normal cells by their typical morphology. The time needed for such a search, however, is considerable, since

the phase contrast (wet mount) preparation fails to give the contrast of colors as found in the Papanicolaou smears. Since prolonged and intensive examination during office-hours is impossible, we prefer to use stained smears for searching for carcinoma cells (the laboratory method). If the Papanicolaou test proves negative, a repeat test is not required for one year. During the interval, check-up examinations may be carried out with the rapid method. Consequently, the time-consuming studies of the cytological laboratory are circumvented.

The use of both methods together represents an ideal of modern medicine. The definitive, rapid, and easily performed method of examination can be applied directly to the patient to provide results which may be used for instituting prompt therapy and results which may be checked by a specialist (cytological laboratory), if such is deemed necessary.

The rapid method (phase-contrast microscopy) is no substitute for the Papanicolaou method (laboratory examination), since:

a) the fresh, wet-mounted preparation cannot be preserved for rechecking. Documentation of the results by photomicrography is too costly.

b) the diagnostic accuracy in cancer detection studies is less than with the Papanicolaou preparation. The hyperchromasia of stained tumor cells makes it easier to recognize them. In phase-contrast preparations the morphology of the cells alone decides how they should be classified. Consequently, more time is needed in wet-mount preparations to find the atypical cells.

The reasons just given, however, do not diminish the value of a method which, if used for orientation, proves exceedingly valuable within the limits described.

A. Preparing the Specimen for Study

1. Obtaining Vaginal Secretions

The specimen of vaginal secretions for the wet-mount studies is taken with a platinum loop under direct vision:

a) for the evaluation of ovarian function — from the upper, lateral vaginal wall.

b) for study of microorganisms — from the upper, lateral vaginal wall.

The same preparation is used for a) and b).

c) for the search for carcinoma cells — from the visible lesion or lesions, — from the portio vaginalis (ectocervix), — and from the endocervical canal.

The secretions in the platinum loop are washed free on a glass slide with a drop of physiological saline. A clean coverslip is placed on the mixture and the preparation is ready for examination.

2. Obtaining Cervical Secretions

After the portio vaginalis (ectocervix) and outer cervical os are sponged and cleansed with a cotton tampon, a platinum loop is inserted into the endocervical canal and secretions collected. These are placed on a clean glass slide, a coverslip is applied, and the preparation is examined.

Preparing the Sims-Huhner Test

Secretions may be collected from the cervical canal up to 12 hours after sexual intercourse. For a positive test the sperm should be active and remain viable for several hours, as their motility will show. The method of study is also suitable for the Miller-Kurzrok Test.

3. Preparing Aspirated Fluids

Fluid is aspirated with a small-gauge needle. A drop of the fluid obtained is promptly examined under the phase-contrast microscope. The remaining fluid is centrifuged, and a drop of the sediment is examined under the microscope; smears are made from the rest of the sediment, fixed and examined according to the Papanicolaou method.

4. Preparing Urine and Urinary Sediment

The mid-stream portion of freshly voided urine is used for study, whereas any portion of urine obtained by catheter is suitable. As with aspirated fluids, a drop of urine is examined promptly, the rest is centrifuged and the sediment studied.

B. Use of Cytology during Clinical Examination

1. The clinical history and menstrual history obtained.
2. The patient on the examination table. Inspection and palpation of the abdomen.
3. The vulva inspected.
4. The vaginal specula inserted. The vagina and portio vaginalis exposed. The vaginal wall and surface of the cervix inspected.
5. Secretions obtained from the upper, lateral, vaginal fornix, and wet-mount smear prepared.
6. Vaginal and cervical secretions obtained for Papanicolaou smears, fixed and sent to the laboratory.
7. Colposcopy.
8. Bimanual vaginal examination.
9. Rectal examination.
10. Examination of the breasts.
11. Study of the wet-mount smear under the phase-contrast microscope (5).
12. Results discussed with the patient.

C. Evaluation of the Wet-Mount Preparation under the Phase-Contrast Microscope

I. Diagnosis of Ovarian Function

Phases of the Menstrual Cycle

With vaginal cytology it should be possible to differentiate between follicular phase (proliferative phase) and the luteal phase (secretory phase), and especially to determine when ovulation will occur. Each phase of the menstrual cycle may be divided further into three groups:

Proliferative Phase

5th—7th day: most of the cells have vesicular nuclei and show little tendency to cluster; moderate numbers of polymorphonuclear leukocytes are present with an occasional erythrocyte.

8th—11th day: increasing numbers of single superficial cells with pyknotic nuclei, but numerous cells with vesicular nuclei still present; few polymorphonuclear leukocytes.

12th—14th day: the superficial cells with pyknotic nuclei clearly predominate. Large, flattened cells and single cells with good cytoplasmic turgor, almost no polymorphonuclear leukocytes.

The most important characteristics of the advancing proliferative phase: large, flattened superficial cells lying singly; few polymorphonuclear leukocytes, a "clean" field (no inflammatory exudate), increasing nuclear pyknosis up until ovulation.

Secretory Phase

15th—17th day: many cells still possess pyknotic nuclei; intermediary cells with vesicular nuclei increase in number. Early signs of rolling-up and folding of cell margins. Increased desquamation with more numerous clustering of cells; increased numbers of polymorphonuclear leukocytes.

18th—24th day: fewer cells with pyknotic nuclei but more numerous intermediate cells with vesicular nuclei. Greater folding of cell margins. Loss of cell turgor, marked clumping of cells.

25th—28th day: rare epithelial cells with pyknotic nuclei; most cells of the intermediary type with vesicular nuclei. Very marked rolling-up and folding of the cell margins. Pronounced clumping of cells. Increasing numbers of polymorphonuclear leukocytes. The smear appears faded and dirty.

The most important characteristics with advancing secretory phase: clumping of cells, rolling-up and folding of cell margins, cells with vesicular nuclei, increasing numbers of polymorphonuclear leukocytes, increasing dirty, untidy appearance of smear.

Menstruation

Dense clumps of epithelial cells, chiefly intermediary cells with vesicular nuclei. Occasional dissolution of cytoplasm. Incorporation of polymorphonuclear leukocytes in the cell clumps. Numerous erythrocytes and histiocytes. Initially, single endometrial cells; later, endometrial cells in clumps.

The Pre-Ovulatory Smear

Single superficial cells with pyknotic nuclei. Cytoplasm flattened out with sharp contours, no polymorphonuclear leukocytes.

The Post-Ovulatory Smear

Superficial cells and intermediary cells, clumping of cells with rolling-up and folding of cell margins, polymorphonuclear leukocytes; occasionally a few erythrocytes directly after ovulation (ovulatory bleeding).

Pregnancy

During pregnancy the effect of progesterone is especially marked. Most epithelial cells are of the intermediary type. The nuclei are vesicular, the margins of the cells are folded, producing the typical navicular cells.
The most common microorganism found during pregnancy is Döderlein's bacillus. Often a Döderlein cytolysis may be seen.
Disturbances during pregnancy produce noticeable changes in the degree of proliferation, corresponding to the criteria of a disturbed pregnancy as seen in the Papanicolaou stain.

Atrophy

When hormonal stimulation is lacking, the vaginal smear is characteristic. It consists of parabasal and basal cells. Most of them stain cyanophilic with the Papanicolaou stain, occasionally light blue. Only rare cells have an eosinophilic cytoplasm. The nuclei show little detail and generally appear uniformly dark (degenerative changes). Some vesicular nuclei may be seen. In general the cellular picture is irregular.
Most cells in the smear lie singly. When their cytoplasmic membranes disintegrate, however, the cells may clump and the irregular, in large part degenerated nuclei may be extruded to coalesce ("atrophic cell-cohesion" of Wied). With the autolysis the cytoplasm may disappear and the structureless nuclei then lie free.
Polymorphonuclear leukocytes abound. The thin vaginal epithelium is easily traumatized, allowing erythrocytes to extravasate. The polymorphonuclear leukocytes and erythrocytes may degenerate. Histiocytes in variable numbers are encountered.
Because glycogen cannot be formed by the parabasal cells, Döderlein's bacilli are unable to flourish. Consequently, the flora of microorganisms is usually mixed.
With inflammatory changes in the vagina, as in senile colpitis, the appearance of the smear becomes so variegated that the abnormal cells which occur suggest the presence of a carcinoma. In such cases it is advisable to induce proliferation with androgens or estrogens, for the resulting build-up of the epithelium to the middle or superficial layer eliminates the difficulty. About 25% of the smears in menopause are of the atrophic type. The percentage increases with the years beyond menopause.

The Androgenic Smear

With partial or complete loss of ovarian function the "third gonad" (adrenal cortex) may produce a proliferative effect. In such cases one refers to an androgenic or adrenal type of proliferation.
Its Characteristics Are: Only intermediary cells or parabasal cells. Generally they lie singly, rarely in clumps. Usually polymorphonuclear leukocytes are present.

The Estrogenic Smear

With a pure estrogenic effect the smear is composed only of superficial cells. These lie singly, are large, flattened out, with pyknotic nuclei. The smear is uniform and appears clean. Polymorphonuclear leukocytes are rare.

II. Diagnosis of Microorganisms

Döderlein's Flora

A pure flora of Döderlein's bacilli is characterized by the uniform appearance of plump, short, or long rods that lie between or on the epithelial cells. Intermediary cells containing glycogen are often covered with the bacteria. With very heavy infection with the Döderlein's bacillus the cytoplasm of the epithelial cells may be lysed, leaving only the nucleus intact; a functional diagnosis becomes impossible.

Trichomonads

These protozoan are distinguished by their typical form and by the beating of their flagella which are readily seen in the fresh, wet-mounted preparation.

Vaginal Mycoses

In typical cases one finds a dense meshwork of filaments (hyphae or mycelium) which are readily recognized by their tubular shape, budding, and segmented branchings. Spores are almost always present.

Hemophilus Vaginalis

These small coccobacilli usually are pleomorphic but may be recognized in heavy infections by the manner in which they coat the epithelial cells, as if these had been powdered with sugar. Usually dense colonies of the bacteria are seen between the cells as well.

Mixed Bacterial Flora

The mixed flora consist of cocci of various forms. Occasional Döderlein's bacilli may be recognized. Staphylococci and streptococci may not always be clearly differentiated by the manner in which each grows. A heavy infection with cocci often causes the cytoplasm of the epithelial cells to lyse. In such instances the nuclei remain intact but a functional diagnosis is impossible.

III. Search for Tumor Cells

Examination in the wet-mount preparations discloses that with the normal maturation of the squamous epithelial cell the nucleus and cytoplasm undergo changes of aging in a coordinated, progressive manner. In the superficial cells as the nucleus shrinks during maturation to become pyknotic specific changes occur in the cytoplasm as well. With increasing age intercellular bridges form; the cell membrane becomes distinct, and coarse particles form in the cytoplasm.

In contrast, in the carcinoma cell maturation is abnormal and the process of aging of its nucleus and cytoplasm becomes disturbed. In immature carcinoma cells the cytoplasm becomes excessively fluid. The cell membrane becomes inapparent and often during examination the cell alters its shape by an ameboid flux of the cytoplasm. In the most anaplastic cells (poorly differentiated epidermoid carcinomas) a naked nucleus with large nucleolus seems to be lost in a sea of poorly organized cytoplasm. In carcinoma cells showing atypical maturation, as in well-differentiated carcinomas and squamous cell carcinomas, the nucleus often appears immature whereas the cytoplasm resembles that of a mature normal cell. These variations in maturation are as important in the diagnosis of carcinoma in phase-contrast microscopy of viable cells as in fixed and stained preparations.

Criteria of Malignancy

Nucleus	**Cytoplasm**
Anisonucleosis	Anisocytosis
Changes in the nuclear-cytoplasmic ratio	
Polymorphism, hyper-, hypo-, and polychromasia.	Polymorphism, basophilia, dissolution of cytoplasm with release of naked nuclei
Atypical chromatine structure	Absence of intercellular bridges
Increase in number and size of nucleolar substance, altered	
nuclear-nucleolar ratio	Phagocytosis (cannibalism)
Mitoses and mitotic disturbances	
Giant nuclei	

Additional criteria

Conglomerates of cells
Presence of leukocytes, erythrocytes, histiocytes

IV. Single Cells

Characteristics of Single Cells in Wet-Mount Preparations

Normal Cells

Basal Cells. These are rounded cells with a distinct cytoplasmic membrane. The large, spherical nucleus usually is central and optically homogeneous. It contains one or two rounded nucleoli which appear dark because of their density. Delicate cytoplasmic granulations, like the occasional perinuclear vacuoles which either lie against the nucleus or surround it, apparently represent structures formed by metabolic changes. Through artifact (pressure on the coverslip, drying) cytoplasm may exude from the cells.

Parabasal Cells. The cell is rounded or oval, the nucleus elongated or circular. The nuclear membrane is distinct, often wrinkled and less smooth than that of the basal cells, and its nuclear substance is denser. The nucleoli are distinct. The cytoplasm contains fine granulations. Occasionally one or more coarse granules is found near the nuclear membrane.

Intermediate Cells. The cells are flattened, especially at the margins; the midportion, however, bulges because of the vesicular nucleus. If the preparation is agitated the cell moves and appears like a disc although irregular in shape. Its cytoplasmic membrane is wrinkled. Intercellular bridges may be evident in some clusters of cells. When the surface of the cell is sharply focused a small ridge may become visible which likewise appears to be a structure important in cell cohesion. The nucleus may be as large as that of the basal cell but its chromatin is denser. Usually the nucleus is smaller than the basal nucleus. Only rarely are one or two nucleoli apparent. The cytoplasm shows increasing granularity; the granules are coarse to fine. They fill up the cytoplasm to the periphery except for a narrow perinuclear halo which they surround irregularly.

Superficial Cells. The superficial cell, the most mature of the squamous epithelium, is extremely flat and scale-like. The granulations are increased and fill the entire cytoplasm.

These cells are often arranged in rows. The cell boundaries are irregular. The points of attachment of the intercellular bridges are indistinct. The nucleus has become a flattened disc and condensed by loss of water. It is usually surrounded by a clear halo (retraction zone, glycogen).

The Anuclear Squama. Anuclear squamae may exfoliate with excessive hornification resulting from local irritation, as in uterine prolapse or in leukoplakia in which the vaginal epithelium becomes thick like the skin. The squamae are as large as the superficial cells but are even flatter and more wrinkled, and have ill-defined margins. The nucleus is either lacking or present only as a ghost. Granulation of the cytoplasm is minimal. Bacteria and leukocytes can often be seen in the grooves and folds of the corrugated surface of the squamae.

Dyskaryotic Cells

As in stained preparations, the following types of cells may be differentiated:

1. Superficial cells showing dyskaryosis.
   Although the cytoplasm resembles in shape and structure that of normal superficial cells, the nucleus does not, for it has failed to become pyknotic. In its size and structure the nucleus is like that of the basal-parabasal cells and is often dense and distorted.
2. Dyskaryotic intermediary cells.
   The cytoplasm reveals the characteristic of the normal intermediary cells with tonofibrils, intercellular bridges and the usual cytoplasmic structure. The nucleus, however, retains the size of nuclei of basal or parabasal cells, although it usually is irregular in shape.
3. Dyskaryotic parabasal cells.
   Since no sharp line of demarcation separates basal and parabasal cells, the diagnosis of a parabasal dyskaryosis cannot be made in the wet-mount preparation.

Carcinoma Cells

The Poorly Differentiated Type. The finely granular nucleus is usually of uniform size although seldom round. It contains one or more enormous nucleoli. The cytoplasm has already disintergrated.

The Basal Type. The cytoplasm shows little structure and is usually plastic, for often during observation the cells may be seen to change their shape (ameboid-like flux). The nuclei resemble those of the poorly differentiated type, although they vary more in shape than those of the poorly differentiated type.

The Pleomorphic Type. When the cytoplasm reaches a certain level of organization through differentiation, then cells of many different types begin to appear. The variation in grade of differentiation is reflected in the pleomorphism of the cells. The cytoplasmic membranes are distinct but the cytoplasmic processes indicate the cell shape is often undergoing change. Cytoplasmic structures are usually prominent, frequently as coarse granules or occasionally as vacuoles. The chromatin of the nuclei varies greatly, from finely granular to coarsely clumped. The nucleoli are large; generally more than one is present in a nucleus. Giant nuclei occur and multinucleated cells can often be found. The nuclei may lie separated within the cell or may, because the cell is compressed, contact one another.

The Cornified Type. These cells originate from the hornifying parts of a squamous cell carcinoma. Their frequency depends on the degree of hornification; that is, on the highly differentiated state of the carcinoma. The cytoplasm is well-organized, not to be distinguished from that of a normal superficial cell. The cell varies in shape, is often rectangular and elongated (spindle cell).

From the criteria given it is possible to classify the types of cells in the following groups:

1. Unimorphic carcinoma cells
   a) poorly differentiated type
   b) basal cell type.
2. Pleomorphic carcinoma cells
   a) pleomorphic type
   b) hornified type.

Columnar Epithelium

The tall columnar epithelium of the endocervical canal consists of ciliated cells and of mucus-secreting cells. The two types can be distinguished in the wet-mount preparations:

1. *Ciliated Cells.* These cells have the shape of blunted cones, with the narrow end the basal part. The flattened, large end is ciliated. Small rounded granules can be visualized at the lower end of the cilia which anchors them in the cytoplasm. The rhythmic beating of the cilia occurs about 60 times a minute. It can be readily observed and ceases only after long periods. The boundaries of the cells are delicate. The large, rounded nucleus is usually located in the midpart of the cell. It has a sharply defined membrane and occasionally contains one or two nucleoli. Secretory droplets may form in the cytoplasm.
2. *Mucus-Secreting Cells.* The columnar cells that produce mucus are plumper and more variable in shape. Often they are rounded in the secretions of freshly mounted preparations. The cytoplasm is usually fluid. Occasionally one finds naked nuclei which can be recognized as nuclei from columnar epithelial cells by their pallisade arrangement.

The columnar cells are unusually delicate and are likely to undergo rapid, secondary changes. The nucleus swells, the cell boundaries become indistinct, and nuclear and cytoplasmic structures fade. As such changes progress, it may become difficult to recognize and classify the cells.

V. Secondary Changes in Cells

The cells in the wet-mounted preparations are unchanged by external influences (fixations, staining). Secondary changes do occur in the cells, however, when preserved in the vaginal secretions too long. The mature cells of the superficial layers, already dehydrated and preserved intravitally, are less involved in secondary changes than are the cells of the deeper layers.

A. Profound changes after prolonged preservation in vaginal secretions occur in:
   1. basal cells and parabasal cells.
   2. endocervical cells.
   3. all stages of atypical cells.
B. Insignificant secondary changes after prolonged preservation in the vaginal secretions occur in:
   1. superficial cells.
   2. intermediary cells.

Both types of cells are secondarily changed by bacterial flora:

bacterial cytolysis (Döderlein's flora),
lysis by pathogenic bacteria (mixed flora).

The cytolysis by the Döderlein's bacillus involves primarily the glycogen-rich intermediary cells.

VI. Artifacts

The following artifacts may occur:

1. Bubbles of air or droplets of oil in the preparation appear as round, sharply delineated, optically empty structures.
2. Remnants of vaginal suppositories: droplets of oil, amorphous debris, crystalline matter, fragments of lint, talcum powder, starch grains.
3. Dehydration of the preparation:
   As cells shrink their indices of refraction also change. With loss of cellular water, "schlieren phenomenon" occurs.
4. Swelling of cells after using hypotonic salt solutions for mounting the smear.

VII. The Inadequate Smear

A smear is inadequate when it precludes interpretation. Such pertains when:

a) much blood obscures the epithelial cells.

b) the inflammatory exudate of a colpitis contains so many polymorphonuclear leukocytes that they make recognition of the epithelial cells difficult.

c) the epithelial cells are destroyed by an excessive growth of Döderlein's bacilli or by pathogenic bacteria (cocci or mixed infections).

In such infections the vaginal secretions must be "cleared up" (by applying estrogen intravaginally, by parenteral injections of estrogen, and by local antibiotic therapy), thus eliminating the pathogenic microorganisms and reestablishing the normal biology of the vagina.

d) by faulty technique too few cells are present in the smear.

e) the smear (freshly prepared, wet-mount) is allowed through neglect to dry out before examination.

II.   Technical Explanations

In addition to the resolving power of the objective and the magnifying power of the eyepiece used, the success of microscopic examination is largely determined by the possibilities of enhancing contrast.

The majority of biological and medical objects are imaged under the microscope with insufficient contrast. Although the light is affected by differences in the refractive index and thickness of individual structures, the resulting phase differences cannot be perceived by the human eye.

The contrast of detailed microscopic structures may be enhanced by either of the following two methods:

I. Modification of the Object

The objects (cells, tissue, etc.) are fixed and stained. These absorb the light waves passing through the specimen to a greater or lesser extent. Owing to intensity differences, high-contrast images are produced.

Drawbacks of this Method

1. The fixing and staining reagents may modify the different structures.

2. The time needed is considerable, preventing examination of the specimen during an operation or office-hours.

3. Unstained specimens and living cells (e.g.), microorganisms) cannot be seen.

II. Modification of the Microscope Light Path

The methods of phase-contrast, interference-contrast, fluorescence and ultraviolet microscopy are increasingly being used for examining unfixed and unstained objects. Above all, phase-contrast microscopy has become firmly adopted as a routine technique in clinical medicine.

Phase-Contrast Method

The light waves coming from an annular diaphragm in the condenser pass through the object; part of the beam is deflected by diffraction. The diffracted rays traverse the objective beside the non-diffracted rays. They are weaker than the direct rays. In addition, the diffracted rays are retarded in phase as compared with the non-diffracted ones. With the majority of biological objects the phase difference amounts to about $^{1}/_{4}$ wavelength.

These beams, which differ both in intensity and in phase, traverse the objective side by side and reach the phase plate located in the rear focal plane of the objective. The annular phase plate affects both the amplitude and the phase of the rays so that in the image plane direct and diffracted rays collide which have roughly identical amplitude and a phase difference of $\lambda/2$ interfere, i.e., wave peaks and troughs of identical amplitude; the result is extinction by interference. The structural details appear darker than the surrounding medium.

In order to satisfy these conditions, the direct and diffracted beams must be separated as clearly as possible. However, with the irregular structures of biological objects there is no complete separation between diffracted and direct light. On the contrary, a certain part of the diffracted light will pass the phase annulus together with the direct light, producing optical artefacts; the different structures are surrounded by more or less pronounced halos (see Fig. 2b).

To make full use of the possibilities inherent in phase-contrast microscopy, it is advisable to use very thin specimens. For cytological work, the cells should be clearly separated and above all not superimposed.

It is also indispensable to adjust the appropriate annular diaphragm in the condenser after every exchange of objectives, i.e., to make it coincide with the annular phase plate in the objective (Fig. 1). Even a minor displacement between the two rings results in a noticeable reduction of contrast. The relative position of the annular diaphragms can be checked with the aid of a centering telescope or with the Optovar magnification changer set to PH.

Due to the special physical conditions of the phase-contrast technique, optimum contrast is obtained with monochromatic light. It has therefore become customary to observe phase objects with a green filter in the light path.

Nomarski Differential Interference Contrast

Like the phase-contrast method, the technique of interference-contrast microscopy suggested by Nomarski allows unstained transparent objects to be reproduced with high contrast.

The light from the lamp field-stop is plane-polarized by means of a polarizer. A birefringent quartz plate consisting of two cemented quartz prisms (a so-called Wollaston prism) splits every light wave into two components. The lateral spacing of the two bundles of rays is only a few microns and remains below the microscope's limit of resolution. The two components of the light wave vibrate perpendicular to each other. After emerging from the condenser, both light waves proceed along parallel paths and converge in the image-side focal plane of the objective. A second Wollaston prism modified according to a suggestion by Prof. Nomarski then recombines the two components. An analyzer crossed at 90° with the transmission direction of the polarizer shifts the two light waves into a common vibration plane so they can interfere.

A phase object moved into the light path changes the phase of the two component waves. Depending on the thickness and refractive index of the various object structures, phase differences are produced between the wavefronts that make the object appear more or less bright against the background.

The Wollaston prism 2, which is also called the principal prism, can be shifted at right angles to the optical axis. In center position it will not affect the path difference of the object. By shifting it to the right or left, a supplementary path difference is introduced in addition to the object's path difference, which makes the background field appear more or less bright. It is thus possible to adapt the contrast to the object structures of interest. If the prism is shifted even further away from the optical axis, the contrast will progressively diminish and finally change to color contrast. Differential interference contrast is distinguished by the following characteristics: the numerical aperture of the objective can be fully utilized. This allows the depth of field to be controlled and sharply defined optical sections to be observed in the specimen. Thus with relatively thick and superimposed objects considerably clearer images may be obtained than is possible by means of phase contrast.

Individual cells or structures have a three-dimensional appearance, similar to oblique bright-field illumination (Fig. 2).

Contrary to conditions in the phase-contrast image no "halo" effect develops. This is of special advantage wherever cells are densely packed or superimposed (see Fig. 63).

Equipment

The Nomarski differential interference-contrast equipment consists of three components and can be used on any Zeiss routine or research microscope, like the phase-contrast equipment. These components are a polarizer, a phase-contrast interference-contrast condenser, and the interference-contrast slide accommodating the analyzer and the principal prism (Fig. 5).

The polarizing filter is inserted in the receptacle above the lamp field-stop in the microscope base and aligned in the east-west direction. The vibration direction of the filter is marked on its mount.

The interference-contrast condenser may be used for bright-field, phase-contrast and Nomarski interference-contrast work. For phase work, the numbers of the annular phase diaphragms in the condenser must agree with those of the phase annuli in the objectives. The auxiliary prisms for interference-contrast work have been computed so that only the $16 \times$ Plan objective and the $16 \times$ Neofluar can be used (auxiliary prism I), or the $40 \times$ Plan objective and the $40 \times$ Neofluar auxiliary prism II) or the $100 \times$ Plan objective (auxiliary prism III).

It is advisable to examine unstained specimens in phase-contrast first. Within seconds the image then may be optically stained by the Nomarski method. For this purpose, the interference-contrast slide and the polarizing filter aligned in the east-west direction are moved into the light path. Depending on the desired magnification, one of the objectives for Nomarski contrast mentioned above is then swung in and the appropriate aux-iliary prism selected in the condenser. Contrast can be controlled by shifting the interference-contrast slide and by opening or closing down the aperture diaphragm.

Microscopes for Gynecological Cyto-Diagnostics

The Zeiss Standard RA microscope with attachment camera and the Zeiss Photomicroscope were used to take the photographs. In principle, any microscope with a centering condenser carrier may be equipped with the phase contrast system. For large clinical laboratories, where there is a frequent demand for photomicrography in addition to visual observation, the new Photomicroscope II (Fig. 3) is particularly well-suited. This instrument has an integral, fully automatic 35 mm camera which ensures consistently optimally exposed 24×36 mm micrographs. All camera functions — from electronic exposure to film advance — are automatically controlled by pressing a single button, allowing the microscopist to devote all his attention to examination of the specimen.

The shortest exposure time is $^1/_{100}$ sec. Experience has shown, however, that the different staining techniques and objective magnifications call for exposures ranging from less than one to several seconds. This would give rise to image motion in the case of rapidly moving objects such as Trichromonadidae, spermatozoa, ciliated epithelia, etc. A microflash unit therefore should be used for this type of work. On all microscopes with built-in illuminator the microflash unit can be easily inserted into the lamp fieldstop insert (see Fig. 3). If the microflash unit is used, the correct exposure time or brightness must be determined with the aid of test exposures.

The microscope most widely used in small laboratories and above all in the daily routine of the medical practitioner is the Standard RA (Fig. 4). This instrument likewise can be used for any conventional observation technique.

Photomicrographic Data

Microscope : Zeiss Standard RA with 35-mm
camera and Zeiss Photomicroscope.

Light source : 6-V 15-W illuminator.
A microflash unit was used for
photographing moving objects
(trichomonads, spermatozoa,
bacteria, vibrating epithelia).
See illustrations.

Objectives : Planachromat, $16\times$, 0.35 N.A., and
Planachromat, $40\times$, 0.65 N.A., for
Nomarski interference-contrast.
Neofluar, $16\times$, 0.40 N.A. "Ph 2",
and Neofluar, $40\times$, 0.75 N.A.
"Ph 2", for phase-contrast.

Film stock : Adox KB 14, KB 17, and Kodak
Plus X Pan 125 ASA (22 DIN).

Filter : Green filter VG 9.

Technical Illustrations

Technische Abbildungen

Illustraciones Técnicales

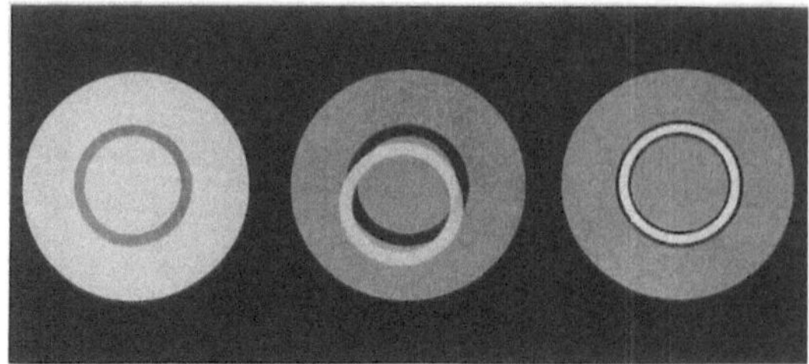

1

Fig. 1. (Left) Phase plate as it appears in the exit pupil of the phase-contrast objective when the annular condenser diaphragm is not in the light path. (Center) With the annular diaphragm of the condenser in the light path, the luminous image of this diaphragm is superimposed on the phase-plate image. (Right) To obtain a perfect phase-contrast effect, the annular diaphragm and the conjugated zone of the phase plate must coincide accurately

Fig. 1. (Links) Phasenplatte, wie sie in der Austrittspupille des Phasenkontrast-Objektives erscheint, wenn die Kondensor-Ringblende nicht eingeschaltet ist. (Mitte) Bei eingeschalteter Kondensorringblende wird das Bild der Phasenplatte vom leuchtenden Bild dieser Ringblende überlagert. (Rechts) Um einen einwandfreien Phasenkontrast-effekt zu erzielen, müssen Ring-blende und konjugierte Zone der Phasenplatte genau zur Deckung gebracht werden

Fig. 1. (Izquierda) Plaquita de fases, tal como aparece en la pupila de salida del objetivo de contraste de fases al no estar intercalado el diafragma anular de condensador. (Centro) Al estar intercalado el diafragma anular de condensador, la imagen luminosa del mismo recubre la imagen de la plaquita de fases. (Derecha) Para obtener un efecto impecable de contraste de fases, es preciso hacer coincidir exacta-mente el diafragma anular y la zona conjugada de la plaquita de fases

Fig. 2. Accessories for Zeiss microscopes for examining objects in Nomarski differential interference contrast, consisting of 1 polarizing filter, 2 combined phase-contrast/differential interference-contrast condenser and 3 interference-contrast slide II for the large research microscopes. For the smaller Standard microscopes, an intermediate tube and the differential interference-contrast slide III are required

Fig. 2. Einrichtung für Zeiss-Mikroskope zur Untersuchung im Differential-Interferenzkontrast nach Nomarski, bestehend aus 1 Polarisations-filter, 2 kombiniertem Phasenkontrast-Differential-Interferenzkontrast-Kondensor sowie 3 Inter-ferenzkontrast-Schieber II für die großen Forschungsmikroskope. Für die kleineren Standard-Mikroskope benötigt man einen Zwischentubus und den Differential-Interferenzkontrast-Schieber III

Fig. 2. Dispositivos correspondientes a micros-copios Zeiss para investigaciones en contraste diferencial de interferencias según Nomarski, compuesto de: 1 Filtro de polarización; 2 Conden-sador combinado de contraste de fases — contraste diferencial de interferencias así como corredera de 3 contraste de interferencias II, para los modelos grandes de microscopios de investigación. Para los microscopios pequeños Standard se necesita un tubo intermedio y la corredera de contraste diferencial de interferencias III

2

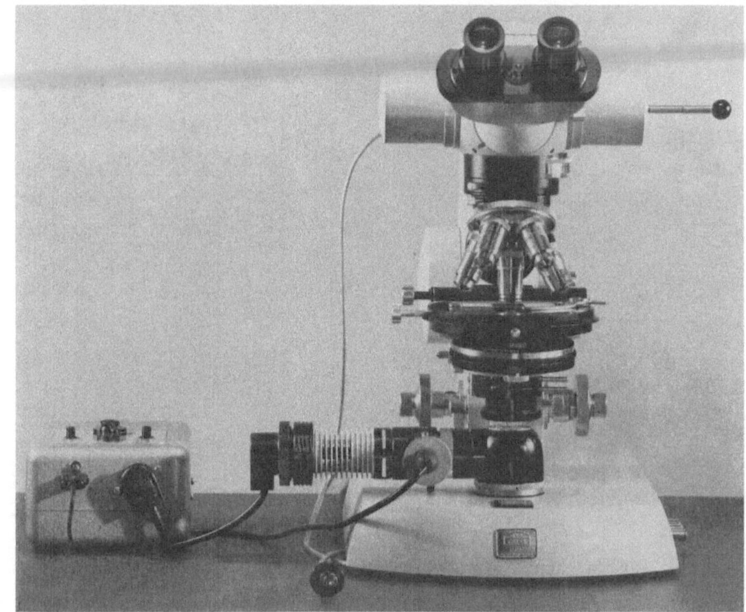

Fig. 3. Photomicro-
scope II with
UN-60 microflash unit

Fig. 3. Photomikro-
skop II mit Mikro-
Blitzgerät UN 60

Fig. 3. Fotomicroscopio
II equipado de micro-
flash UN 60

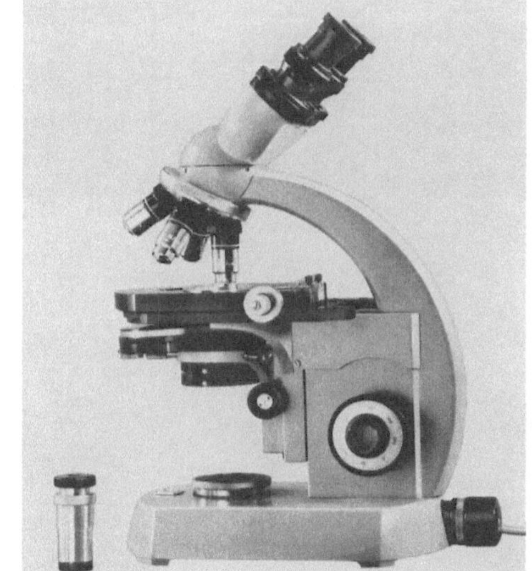

Fig. 4. Standard RA Routine and Research
Microscope, equipped for transmitted-light
phase-contrast work

Fig. 4. Arbeits- und Forschungsmikroskop
Standard RA, ausgerüstet für Unter-
suchungen im Durchlicht-Phasenkontrast

Fig. 4. Microscopio de trabajo e investigación Standard RA equipado para efectuar
investigaciones en luz transmitida y
contraste de fases

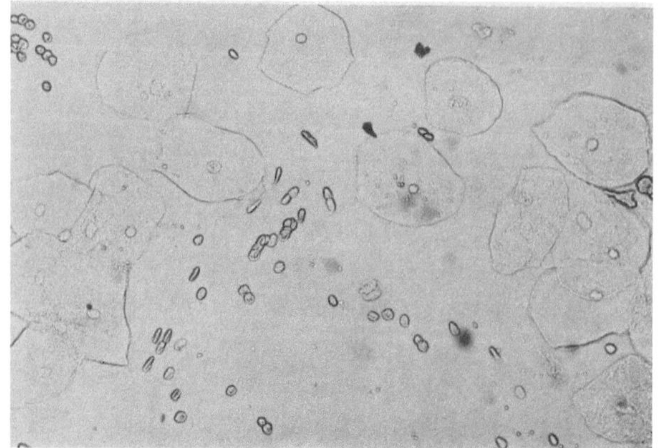

Fig. 5a
Bright field

Fig. 5a
Hellfeld

Fig. 5a
Campo claro

5a

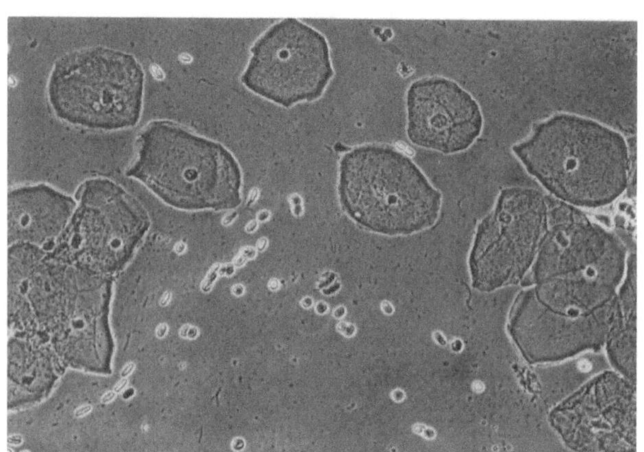

Fig. 5b
Phase contrast

Fig. 5b
Phasenkontrast

Fig. 5b
Contraste de fases

5b

Fig. 5c
Oblique bright field

Fig. 5c
Schräges Hellfeld

Fig. 5c
Campo claro oblicuo

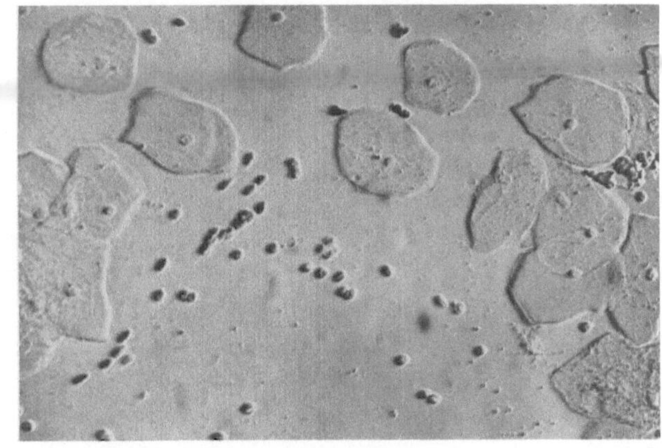

5c

Fig. 5d
Nomarski interference contrast

Fig. 5d
Interferenzkontrast Nomarski

Fig. 5d
Contraste de interferencias Nomarski

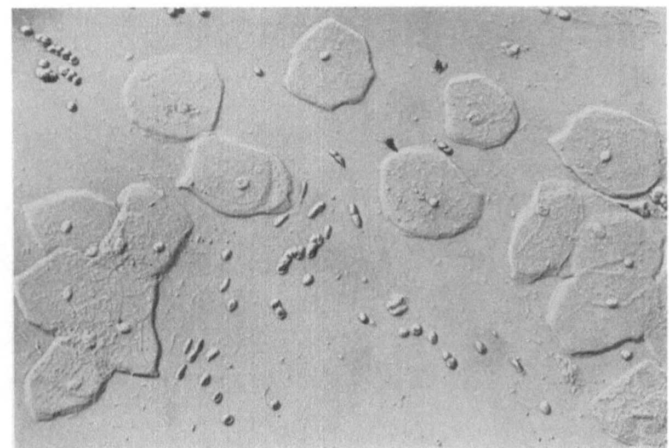

5d

III. Cytological Illustrations

Function · Microbiology · Neoplasia

Cytologische Abbildungen

Funktion · Mikrobiologie · Neoplasie

Illustraciones Citológicas

Función · Microbiología · Neoplasia

The Menstrual Cycle

Cyclus

Ciclo

1. Menstruation
   Adhering clumps
   of endometrial cells;
   a few erythrocytes and
   white blood cells are present

1. Menstruation
   Zusammenhängende Haufen
   von Endometriumzellen.
   Einige Erythrocyten und
   Leukocyten

1. Menstruación
   Grupos de células
   endometriales aglomeradas.
   Algunos eritrocitos y
   leucocitos

2. Fifth day of cycle
   The vaginal epithelium at
   this time begins to proliferate.
   The cells lie singly, have
   vesicular nuclei; no nuclear
   pyknosis. Numerous
   polymorphonuclear leukocytes, some cellular detritus,
   two rounded parabasal cells

2. 5. Cyclustag
   Beginnende Proliferation des
   Vaginalepithels. Die Zellen
   liegen einzeln, bläschen-
   förmiger Kern, noch keine
   Kernpyknose. Zahlreiche
   Leukocyten, etwas Zelldetritus,
   zwei Parabasalzellen

2. 5° día del ciclo
   Comienzo de la proliferación
   del epitelio vaginal. Las células
   yacen aisladas, núcleos de
   aspecto vesicular, picnosis
   nuclear todavía ausente.
   Abundantes leucocitos,
   algo de detritus celular,
   dos células parabasales

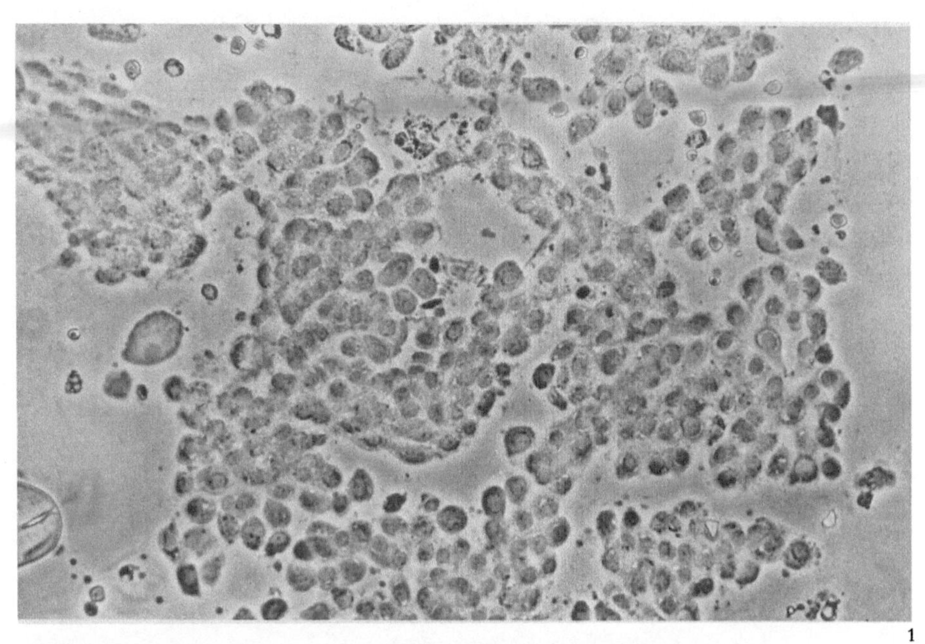

1

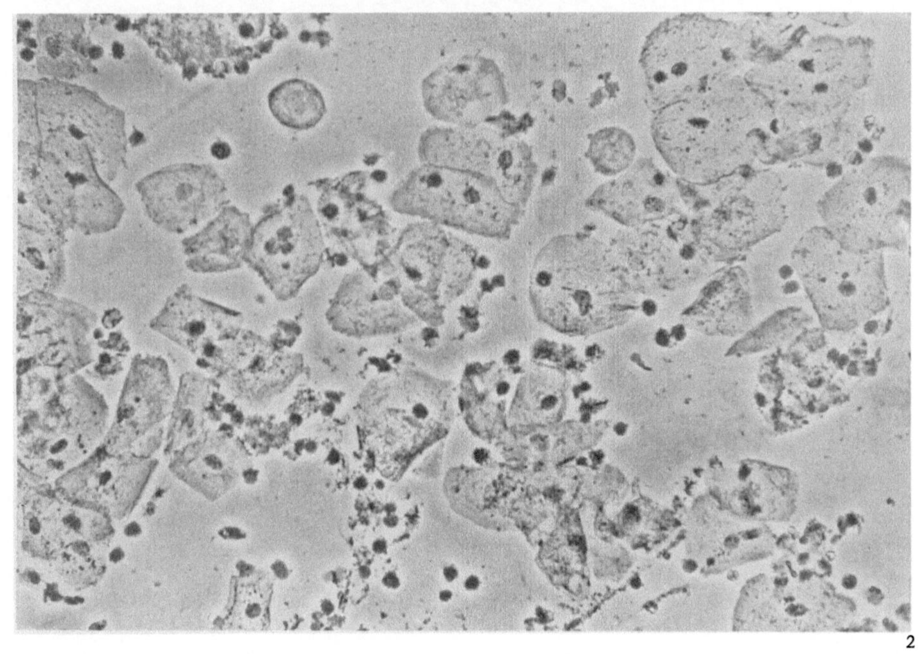

2

The Menstrual Cycle

3. Seventh day of cycle

Increasing estrogen effect. The cells segregated, some have pyknotic nuclei. Erythrocytes are rare, a few polymorphonuclear leukocytes

4. Twelvth day of cycle

(Preovulatory smear) Heightened estrogen effect, the cells lie separated, are flattened out and have pyknotic nuclei. Polymorphonuclear leukocytes are rare

Cyclus

3. 7. Cyclustag

Zunehmender Oestrogeneffekt. Einzeln liegende Zellen, stellenweise bereits pyknotischer Kern. Vereinzelte Erythrocyten, wenig Leukocyten

4. 12. Cyclustag

(Präovulatorischer Ausstrich) Hoher Oestrogeneffekt, einzeln liegende Zellen mit pyknotischem Kern, die Zellen sind flach ausgebreitet, nur vereinzelte Leukocyten

Ciclo

3. 7º día del ciclo

Creciente efecto estrogénico. Células aisladas cuyos núcleos tienden a la picnosis. Eritrocitos aislados, pocos leucocitos

4. 12º día del ciclo

(Frotis preovulatorio) Elevado efecto estrogénico, las células se presentan aisladas, de aspecto plano y extendido, con núcleos picnóticos. Sólo uno que otro leucocito

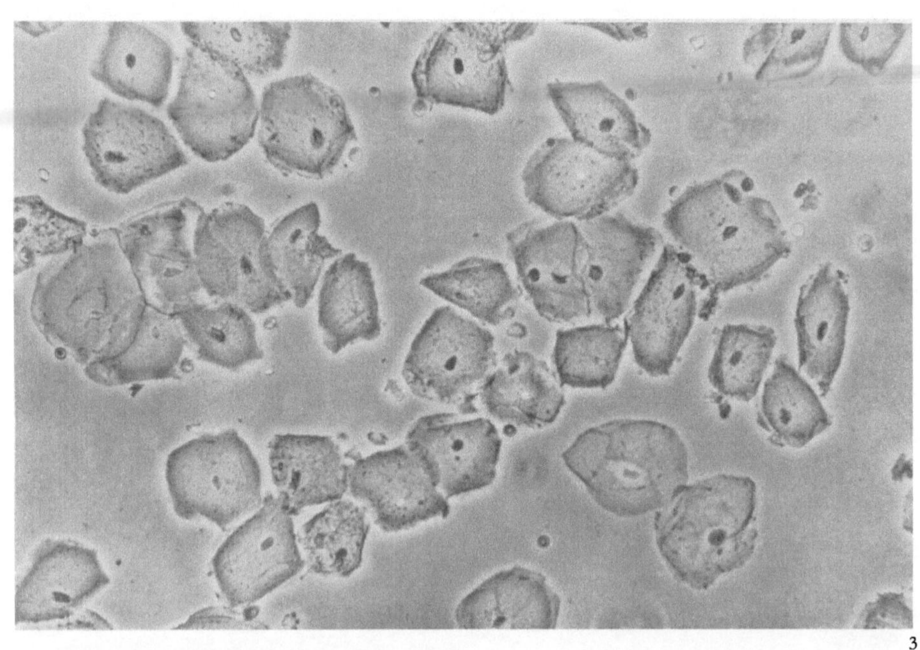

3

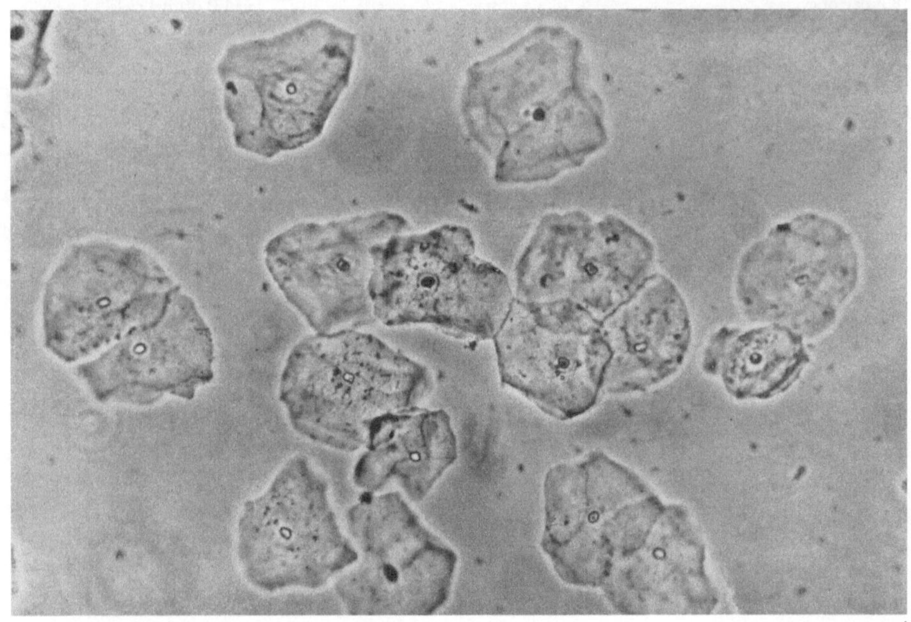

4

The Menstrual Cycle / Cyclus / Ciclo

5. 15th day of cycle
(Postovulatory smear)
The effect of estrogen still evident, the cells have begun to coalesce, rare pyknotic nuclei, cytoplasmic membrane begins to show wrinkles and folds. A few erythrocytes (ovulatory bleeding)

5. 15. Cyclustag
(Postovulatorischer Ausstrich)
Oestrogeneffekt noch erkennbar, aber bereits Zusammentreten der Zellen, nur noch vereinzelte pyknotische Kerne, beginnende Auffaltung des Cytoplasmas, einzelne Erythrocyten (Ovulationsblutung)

5. 15° día del ciclo
(Frotis postovulatorio)
Efecto estrogénico todavía reconocible. Las células comienzan a reagruparse, sólo algunos núcleos picnóticos todavía. Arrugamiento inicial del citoplasma, eritrocitos aislados (sangre de la ovulación)

6. 18th day of cycle
Cells have aggregated, only intermediary cells present. The margins of all are folded. Beginnung regression due to progesterone

6. 18. Cyclustag
Haufenbildung, ausschließlich Intermediärzellen. Alle Zellen zeigen Auffaltung ihrer Ränder (beginnende gestagene Regression)

6. 18° día del ciclo
Aglomeración celular, exclusivamente células intermedias. Todas las células muestran arrugamiento de sus bordes. (Regresión progestacional inicial)

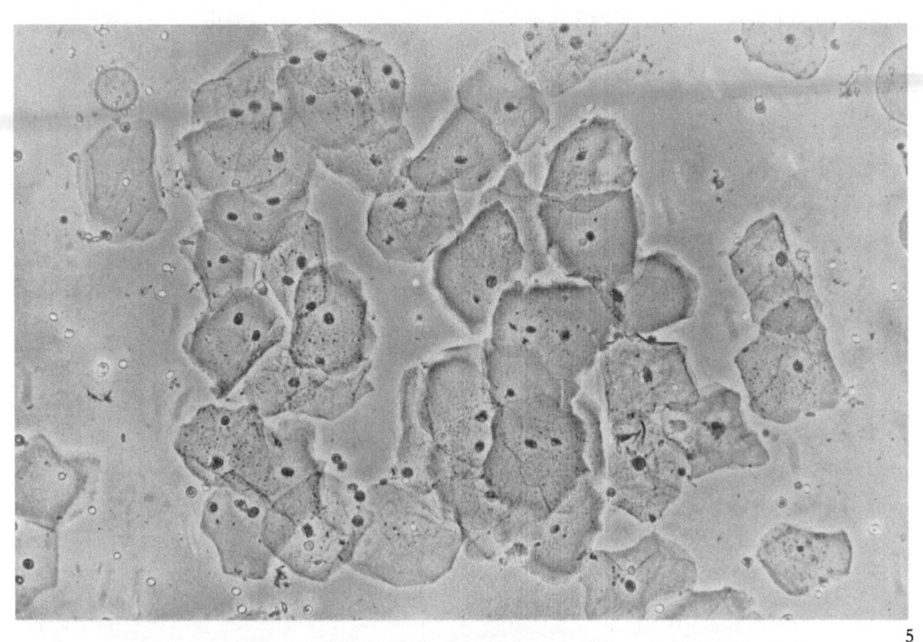

5

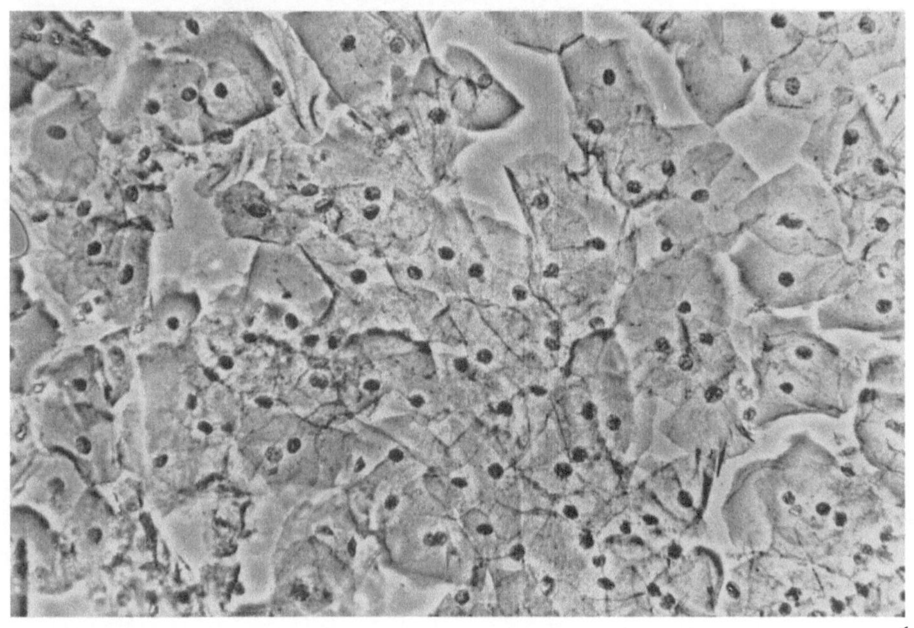

6

The Menstrual Cycle

7. 22nd day of cycle

Formation of cell-aggregates,
only intermediary cells present
with folded margins

Cyclus

7. 22. Cyclustag

Haufenbildung, ausschließlich
Intermediärzellen mit aufge-
falteten Rändern

Ciclo

7. 22° día del ciclo

Exclusivamente células
intermedias, aglomeradas, de
bordes arrugados

8. 25th day of cycle

Intermediate cells with
markedly folded margins,
clumping of cells

8. 25. Cyclustag

Ausschließlich Intermediärzellen, starke Auffaltung der
Ränder, Haufenbildung

8. 25° día del ciclo

Solamente células intermedias,
marcado plegamiento de los
bordes, formación de
conglomerados celulares

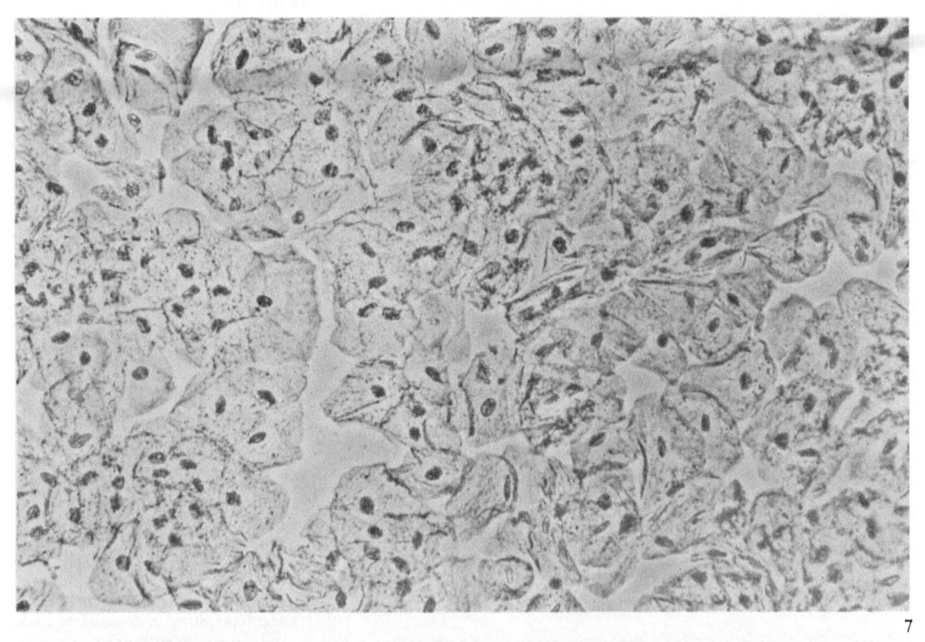

7

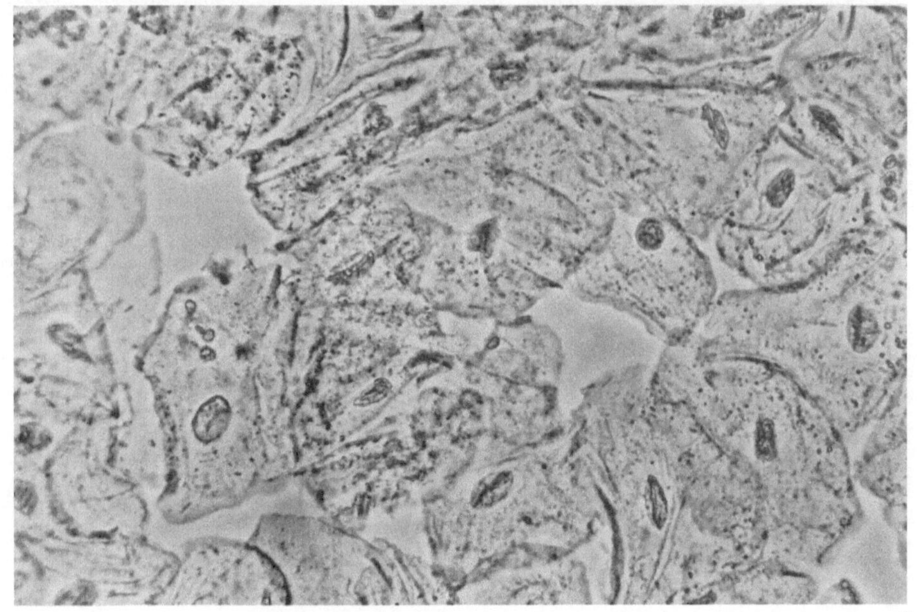

8

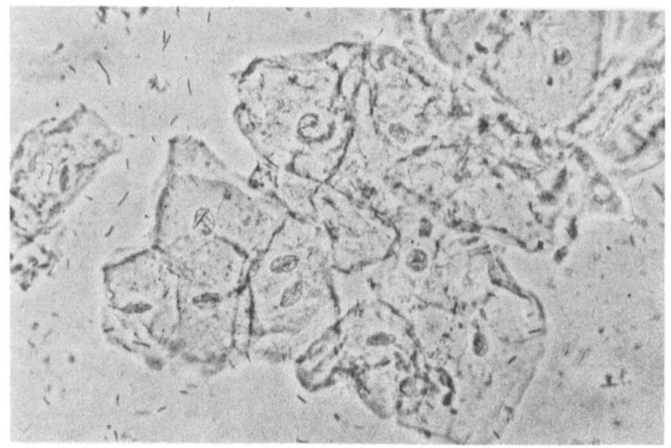

9

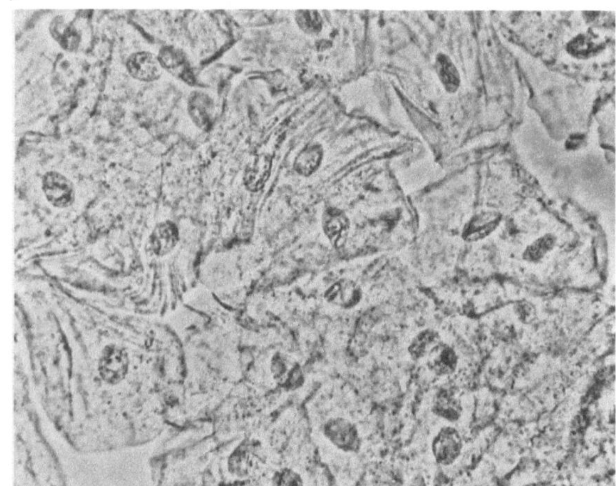

10

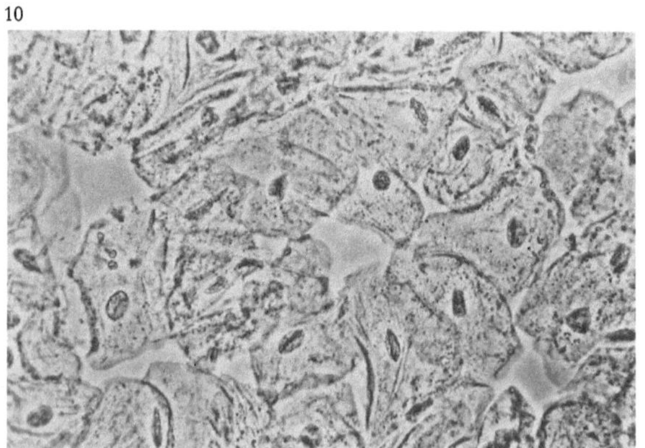

11

Pregnancy

9. Between 1st and 2nd month of pregnancy
   Intermediary cells with vesicular nuclei, cells aggregated, Döderlein's bacilli

10. Between 1st and 2nd month of pregnancy
    Menstruation eight days overdue. Clumping of intermediary cells with rolled and folded margins (navicular type)

11. Fourth month of pregnancy
    Clumping of intermediary cells of the navicular type

Gravidität

9. Gravidität mens I/II
   Intermediärzellen mit bläschenförmigem Kern, Haufenbildung, Döderleinflora

10. Gravidität mens I/II
    Menstruation seit 8 Tagen überfällig, Haufenbildung intermediärer Zellen mit aufgefalteten Rändern (Naviculartyp)

11. Gravidität mens IV
    Ausschließlich Intermediärzellen vom Naviculartyp, Haufenbildung

Gestación

9. 1º y 2º mes de gestación
   Células intermedias con núcleos de aspecto vesicular, aglomeradas, flora de Döderlein

10. 1º y 2º mes de gestación
    Atraso mestrual de 8 días. Células intermedias, aglomeradas, de bordes arrugados (tipo navicular)

11. 4º mes de gestación
    Conglomerados celulares. Exclusivamente células intermedias de tipo navicular

Pregnancy

12. Fourth month of pregnancy

    Copious Döderlein flora. The cyto-
    plasm of the epithelial cells is under-
    going lysis by the Döderlein's bacilli,
    only vesicular nuclei remain

13. Fourth month of pregnancy

    Colpitis, most of the cells shown are
    intermediary cells; the infecting
    bacteria are Hemophilus vaginalis
    wich cause lysis of some cells.
    Bacteria are numerous, some erythro-
    cytes are present

14. Ninth month of pregnancy

    Clumping of intermediary cells,
    Döderlein's bacilli, growth of fungal
    hyphae on the clumped cells (Thrush
    or Candidiasis of pregnancy)

Gravidität

12. Gravidität mens IV

    Ausgeprägte Döderleinflora mit
    Döderleincytolyse. Das Cytoplasma
    ist aufgelöst, es bleiben nur die
    bläschenförmigen Kerne erkennbar

13. Gravidität mens IV

    Kolpitis, vorwiegend Intermediär-
    zellen, bakterielle Verunreinigung
    durch Hämophilus vaginalis. Stellen-
    weise bakterielle Autolyse. Bakterien-
    rasen, einige Erythrocyten

14. Gravidität mens IX

    Intermediärzellen, Haufenbildung,
    Döderleinkeime. Den Zellhaufen
    aufgelagert sind Pilzfäden (Soor-
    mykose in der Gravidität)

Gestación

12. 4º mes de gestación

    Gran abundancia de flora y citolisis
    de Döderlein. El citoplasma está
    disuelto, permaneciendo sólo los núcleos
    de aspecto vesicular reconocibles

13. 4º mes de gestación

    Colpitis. Casi sólo células intermedias.
    Enturbiaminto bacteriano por la
    presencia de Hemophilus vaginalis.
    En algunos sectores autolisis celular.
    Extendidos leucocitarios, algunos
    eritrocitos

14. 9º mes de gestación

    Células intermedias, aglomeración
    celular, bacilos de Döderlein.
    Adosados a los conglomerados
    celulares hay filamentos de hongos
    (micosis en el embarazo)

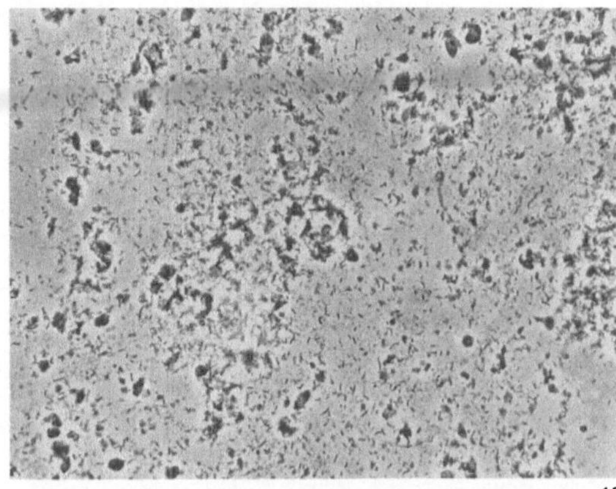

12

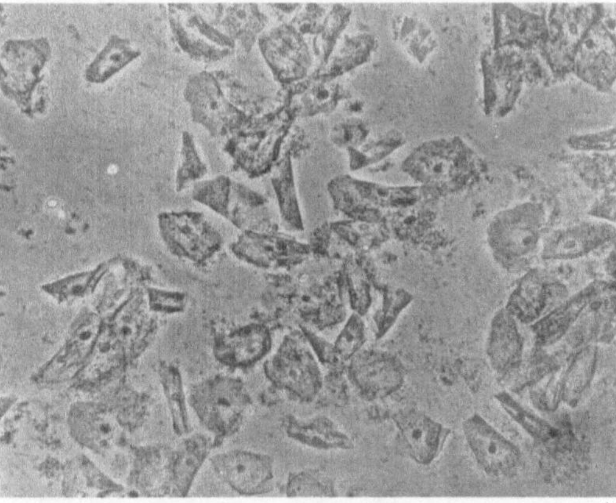

13

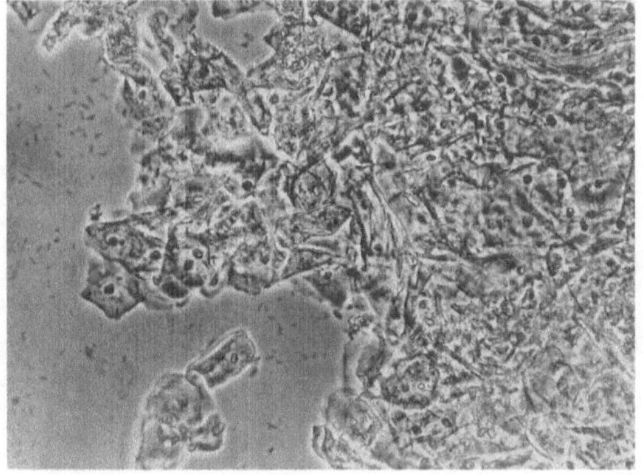

14

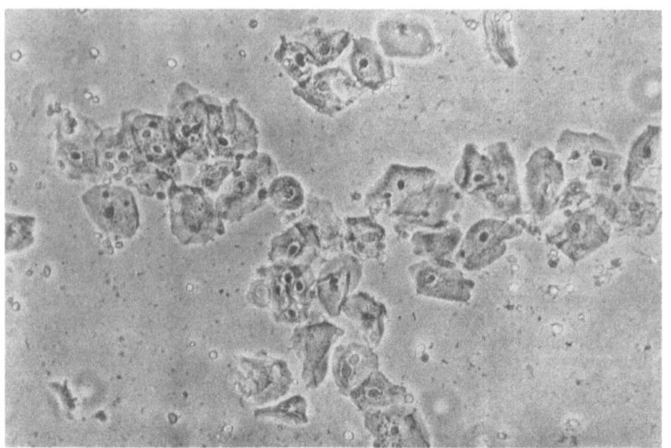

15

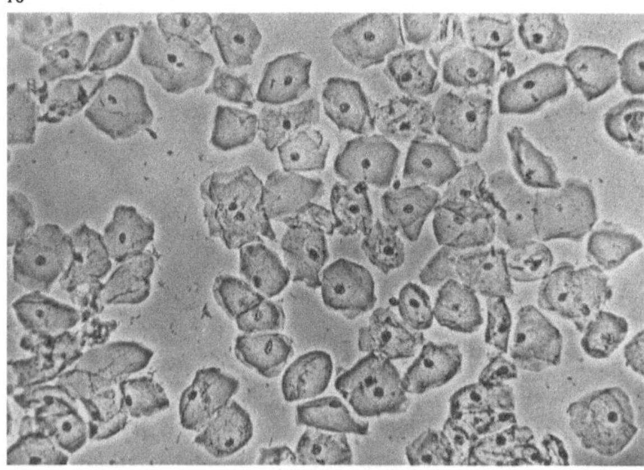

16

17

Smear Showing Androgen Effect

15. Minimal androgenic effect
    Mostly intermediary and parabasal cells, numerous polymorphonuclear leukocytes, an occasional erythrocyte

16. Good androgenic effect
    Segregated, moderately large intermediate cells, polymorphonuclear leukocytes, occasional erythrocytes

17. Good androgenic effect
    Intermediary cells lying singly, Döderlein flora

Androgener Ausstrich

15. Geringer Androgeneffekt
    Vorwiegend Intermediärzellen und Parabasalzellen, zahlreiche Leukocyten, vereinzelte Erythrocyten

16. Guter Androgeneffekt
    Einzeln liegende, mäßig große Intermediärzellen, Leukocyten, vereinzelte Erythrocyten

17. Guter Androgeneffekt
    Einzeln liegende Intermediärzellen, Döderleinflora

Frotis androgénico

15. Bajo efecto androgénico
    Casi sólo células intermedias y parabasales, abundantes leucocitos, algunos eritrocitos

16. Buen efecto androgénico
    Células intermedias aisladas, de tamaño algo mayor que mediano, leucocitos, algunos eritrocitos

17. Buen efecto androgénico
    Células intermedias aisladas, flora de Döderlein

Smear Showing Androgen Effect

18. Smear showing androgenic effect
    Intermediary cells and large type of parabasal cells

19. Androgenic effect
    Intermediary cells and single parabasal cells

20. Androgenic effect and colpitis
    Intermediary cells and parabasal cells, numerous polymorphonuclear leukocytes, occasional erythrocytes. Because of the inflammation the smear appears blurred

Androgener Ausstrich

18. Androgener Ausstrich
    Intermediärzellen und große Parabasalzellen

19. Androgeneffekt
    Intermediärzellen und vereinzelte Parabasalzellen

20. Androgeneffekt und Kolpitis
    Intermediärzellen und Parabasalzellen, zahlreiche Leukocyten, vereinzelte Erythrocyten. Durch die Entzündung macht das Bild einen verwaschenen Eindruck

Frotis androgénico

18. Frotis androgénico
    Células intermedias y grandes células parabasales

19. Efecto androgénico
    Células intermedias y algunas células parabasales

20. Efecto androgénico y colpitis
    Células intermedias y parabasales, abundantes leucocitos, escasos eritrocitos. Debido a la inflamación que acompaña al cuadro, éste impresiona como enturbiado

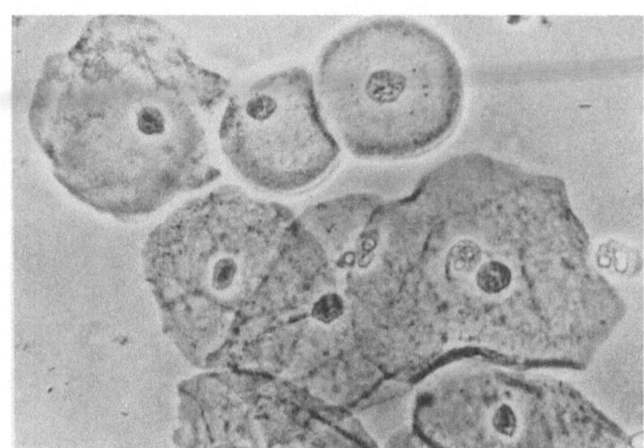

18

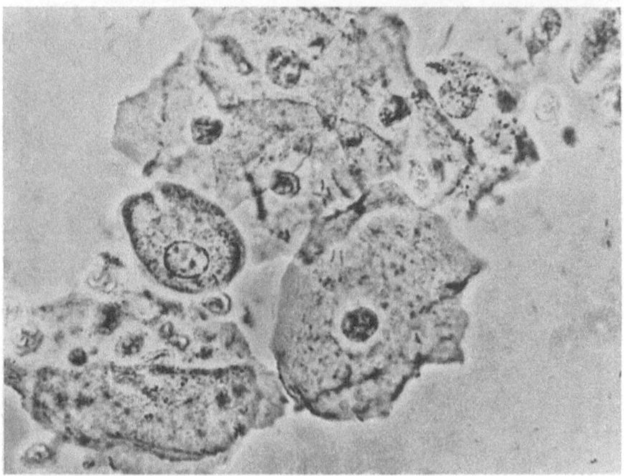

19

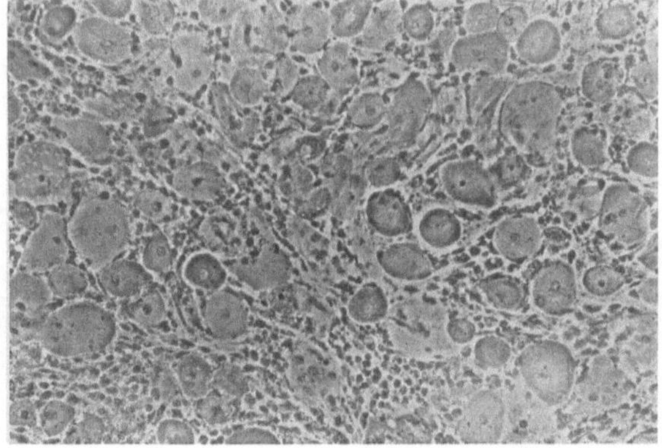

20

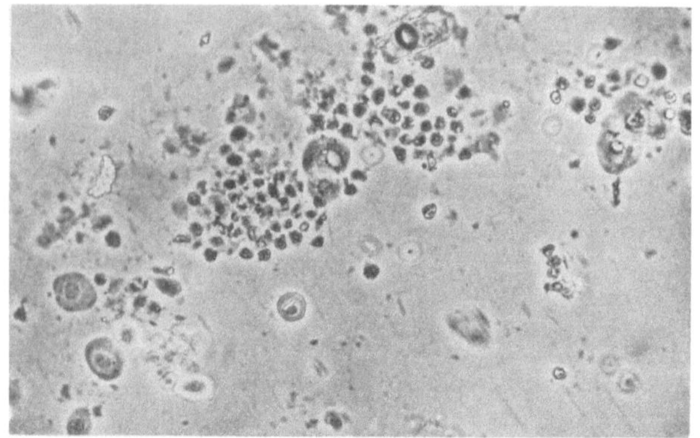

21

Atrophy

21. Smear of atrophic epithelium
    with colpitis
    Exclusively parabasal cells,
    numerous polymorphonuclear
    leukocytes

22. Smear of atrophic epithelium
    Chiefly parabasal cells and inter-
    mediary cells, countless poly-
    morphonuclear leukocytes

23. Smear of atrophic epithelium
    Minimal proliferation with para-
    basal cells and a few intermediary
    cells. Some polymorphonuclear
    leukocytes and erythrocytes

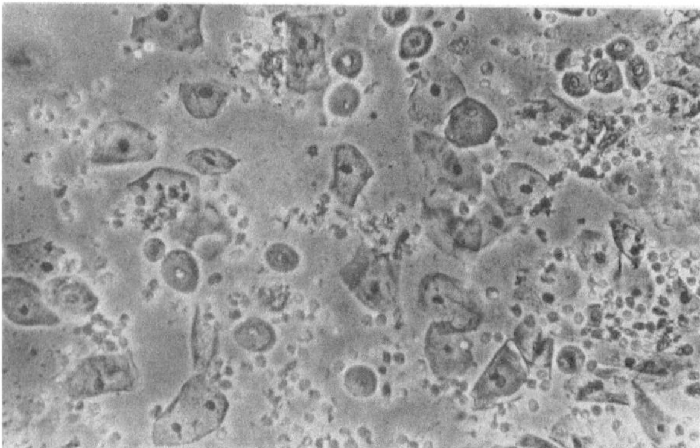

22

Atrophie

21. Atrophischer Ausstrich
    mit Kolpitis
    Ausschließlich Parabasalzellen,
    zahlreiche Leukocyten

22. Atrophischer Ausstrich
    Vorwiegend Parabasalzellen und
    Intermediärzellen, zahlreiche
    Leukocyten

23. Atrophischer Ausstrich
    Niedrige Proliferation mit Para-
    basalzellen und einzelnen Inter-
    mediärzellen. Leukocyten, einige
    Erythrocyten

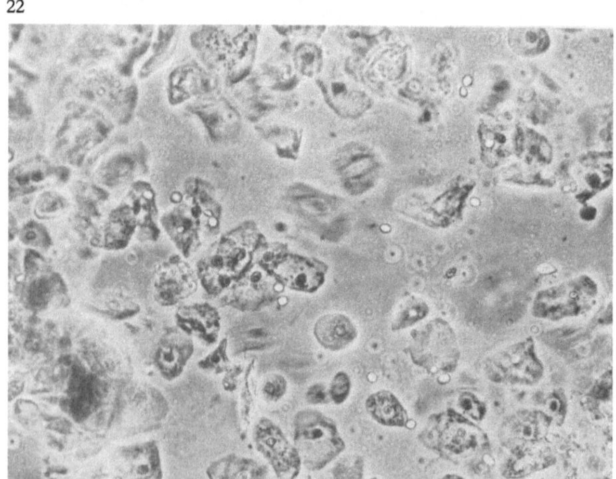

23

Atrofia

21. Frotis atrófico con colpitis
    Exclusivamente células parabasales,
    abundantes leucocitos

22. Frotis atrófico
    Casi sólo células parabasales e inter-
    medias, abundantes leucocitos

23. Frotis atrófico
    Escasa actividad proliferativa con
    células parabasales y algunas inter-
    medias. Leucocitos y escasos
    eritrocitos

Atrophy

24. Smear of atrophic epithelium

    Primarily parabasal cells but also some
    intermediary cells. A hint that atrophic
    cells are beginning to cohere

25. Smear from atrophic vaginal
    epithelium

    An aggregate of parabasal cells,
    cohesion of atrophic cells

26. Smear from atrophic vaginal
    epithelium

    Chiefly parabasal cells and a few
    intermediary cells lying singly

Atrophie

24. Atrophischer Ausstrich

    Vorwiegend Parabasalzellen, auch
    einige Intermediärzellen. Angedeutete
    atrophische Zellkohäsion

25. Atrophischer Ausstrich

    Gruppe von Parabasalzellen,
    atrophische Zellkohäsion

26. Atrophischer Ausstrich

    Vorwiegend Parabasalzellen, einzeln
    liegend, vereinzelte Intermediärzellen

Atrofia

24. Frotis atrófico

    Predominantemente células parabasales.
    Cohesión célular de la atrofia inicial

25. Frotis atrófico

    Grupo de células parabasales. Cohesión
    celular de la atrofia

26. Frotis atrófico

    Predominan células parabasales
    colocadas aisladamente, muy escasas
    células intermedias

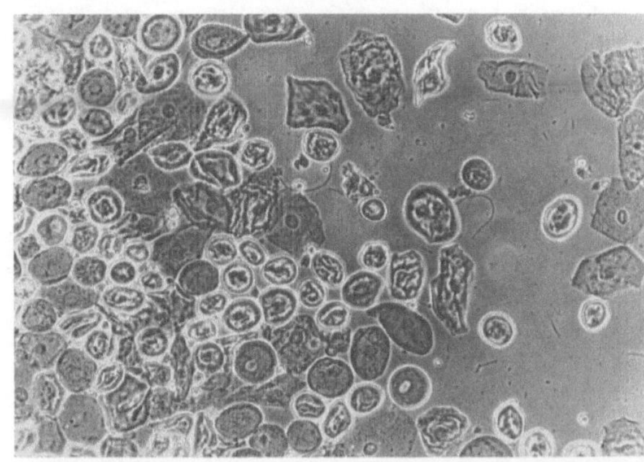

24

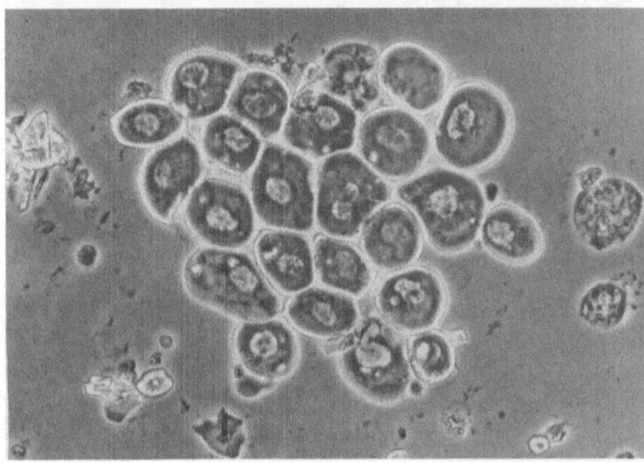

25

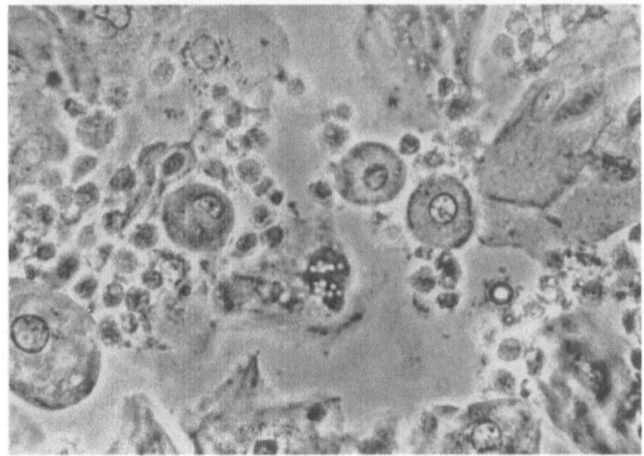

26

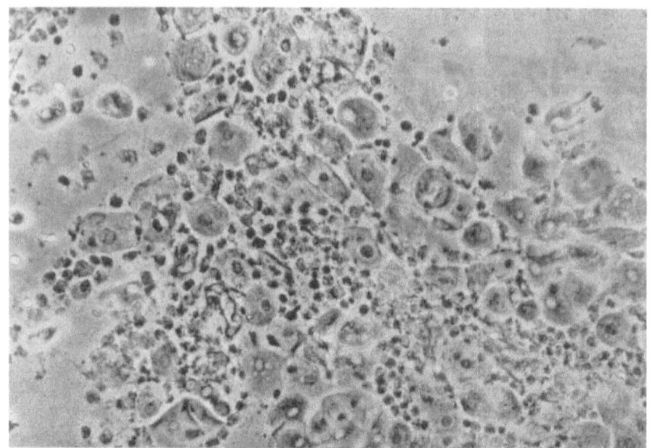

27

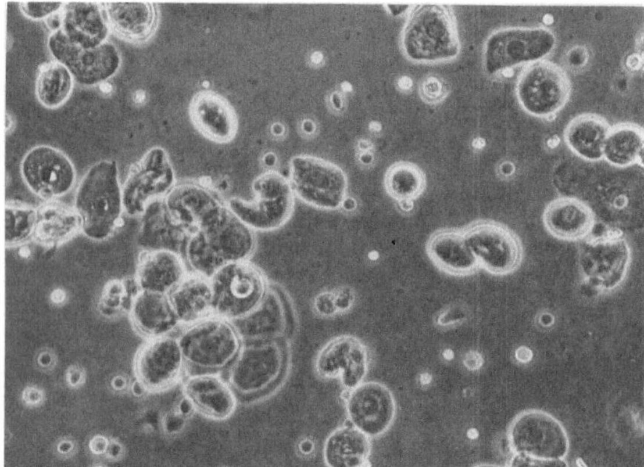

28

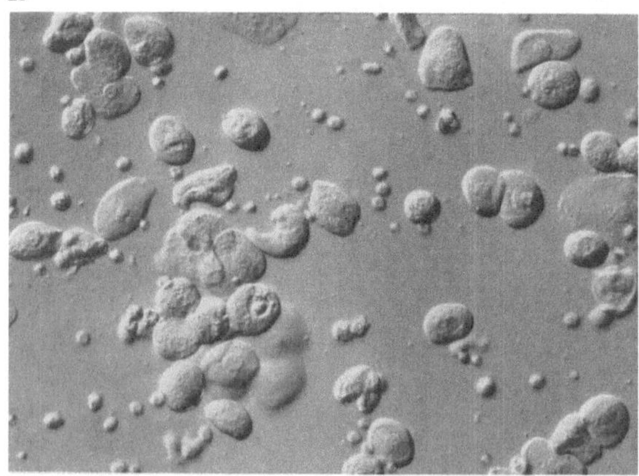

29

Atrophy

27. Smear of atrophic cells with many polymorophonuclear leukocytes
28. Smear of atrophic cells. Phase-contrast
29. Interference-contrast of Fig. 28

Atrophie

27. Atrophischer Ausstrich mit vielen Leukocyten
28. Atrophischer Ausstrich. Phasen-kontrast
29. Interferenzkontrast von Fig. 28

Atrofia

27. Frotis atrófico con muchos leucocitos
28. Frotis atrófico. Contraste de fases
29. contraste de interferencia de la Fig. 28

Atrophy

30. Parabasal cells, a few intermediary cells and trichomonads

31. Cohesion of atrophic cells. Phase-contrast

32. Interference-contrast of Fig. 31

Atrophie

30. Parabasalzellen, einige Intermediär-zellen, daneben Trichomonaden

31. Atrophische Zellkohäsion. Phasen-kontrast

32. Interferenzkontrast von Fig. 31

Atrofia

30. Células parabasales, algunas inter-medias, en las cercanías algunas tricomonas

31. Cohesión celular de la atrofia. Contraste de fases

32. contraste de interferencia de la Fig. 31

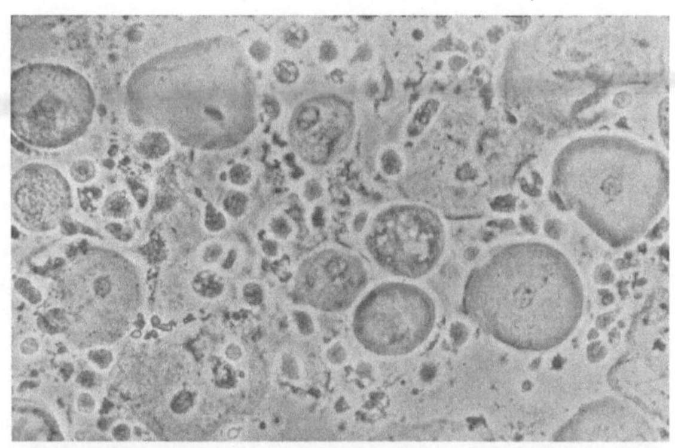

30

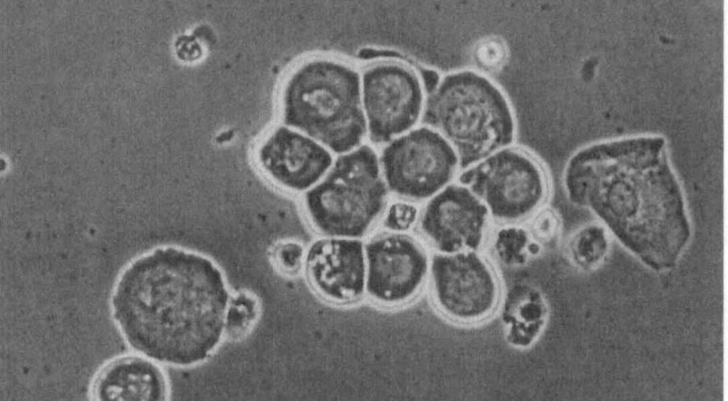

31

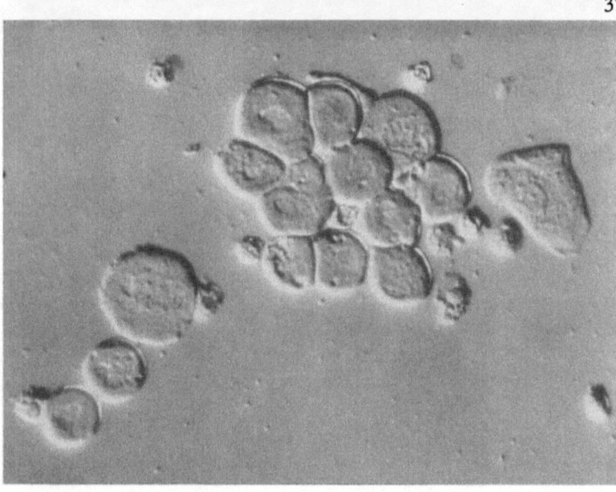

32

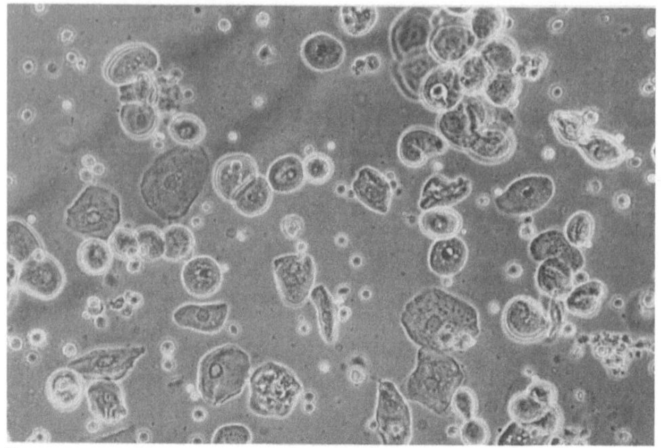

33

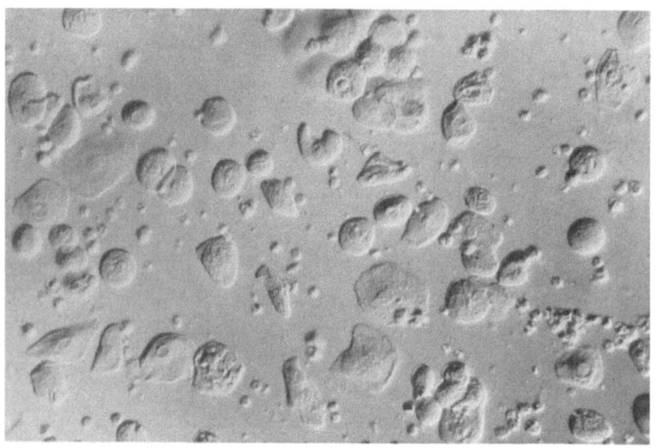

34

Hormonal Effect

33. and 34. Smear of atrophic
epithelium from an old
woman
(Phase-contrast and interference-contrast)
Chiefly parabasal cells and a
few intermediary cells;
abundant polymorpho-
nuclear leukocytes

Hormoneffekt

33. und 34. Atrophischer Ausstrich aus dem Senium
(Phasenkontrast und Interferenzkontrast)
Vorwiegend Parabasalzellen,
daneben auch einige Intermediärzellen; reichlich
Leukocyten

Efecto hormonal

33. y 34. Frotis atrófico de la
senilidad
(contraste de fase e interferencia):
Casi sólo células parabasales,
entre ellas también algunas
células intermedias.
Abundantes leucocitos

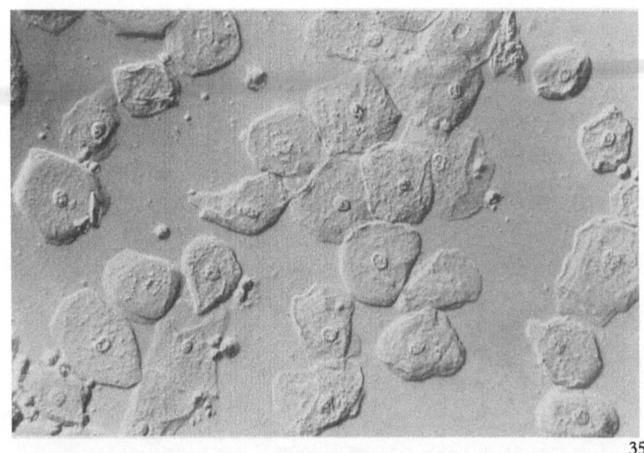

35

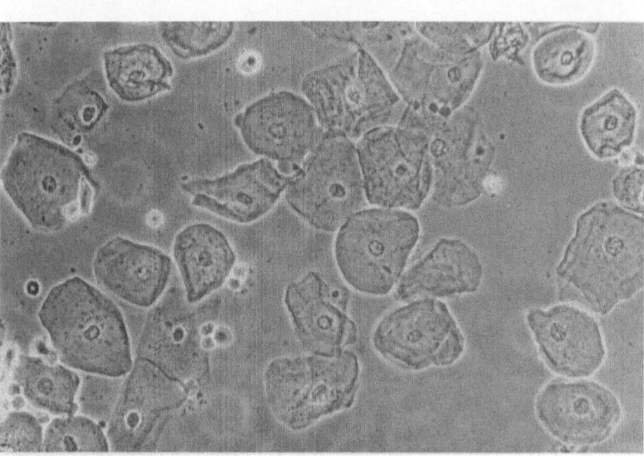

36

Hormonal Effect

35. and 36. Eight days after
therapy with 100 mg of
testosterone proprionate
(Phase-contrast and interference-contrast);

moderate androgenic
proliferation

Hormoneffekt

35. und 36. 8 Tage nach Zufuhr
von 100 mg Testosteron-
Propionat
(Phasenkontrast und Interferenzkontrast)

Mittlere androgene Proli-
feration

Efecto hormonal

35. y 36. 8 días después de la
administración de 100 mg. de
propionato de Testosterona
(contraste de fase e interferencia)

Proliferación androgénica

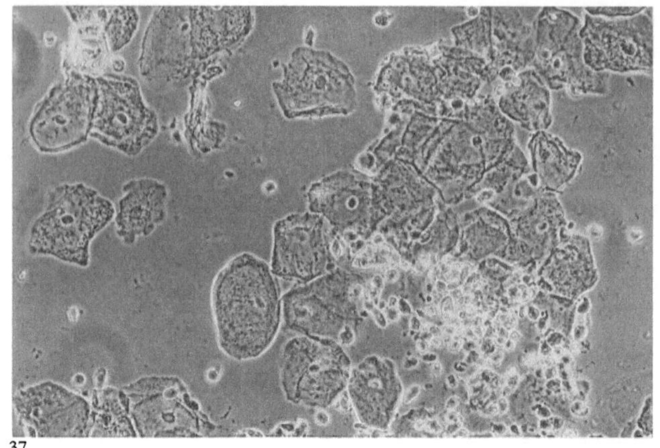

37

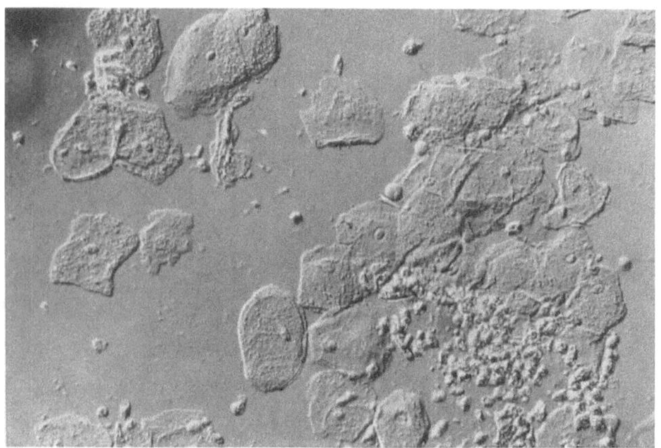

38

Hormonal Effect

37. and 38. Twenty-four hours
after therapy with 100 mg of
progesterone
(Phase-contrast and interference-contrast):

distinct progesterone effect

Hormoneffekt

37. und 38. 24 Std nach Zufuhr
von 100 mg Progesteron
(Phasenkontrast und Interferenzkontrast):
Deutlicher Gestageneffekt

Efecto hormonal

37. y 38. 24 horas después de la
administración de 100 mg. de
Progesterona
(contraste de fase e interferencia):
Evidente efecto secretorio

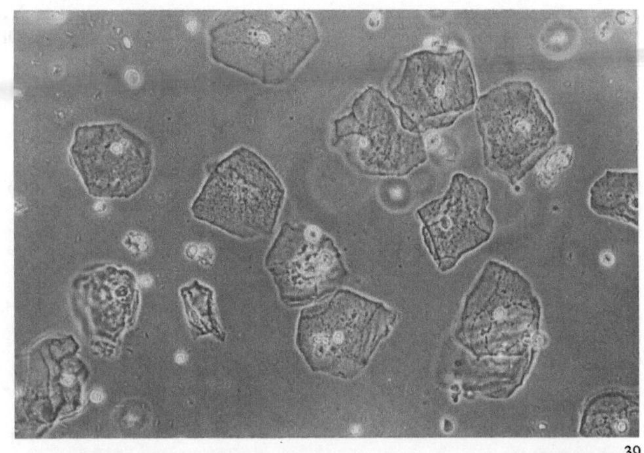

39

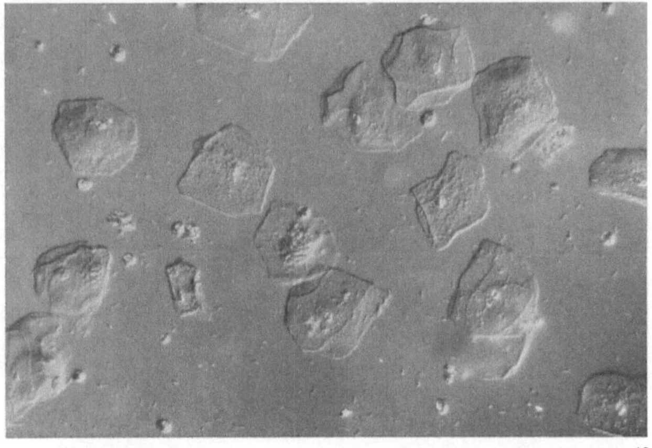

40

Hormonal Effect

39. and 40. Forty-eight hours
after therapy with 10 mg of
estradiol benzoate
(Phase-contrast and interference-contrast):
cells lie singly, nuclear
pyknosis, distinct estrogenic
effect

Hormoneffekt

39. und 40. 48 Std nach Zufuhr
von 10 mg Oestradiol-
Benzoat
(Phasenkontrast und Inter-
ferenzkontrast):
Einzeln liegende Zellen,
Kernpyknose. Deutlicher
Oestrogeneffekt

Efecto hormonal

39. y 40. 48 horas después de la
administración de 10 mg. de
benzoato de Estradiol
(contraste de fase e interferencia):
Células dispuestas aislada-
mente, picnosis nuclear.
Evidente efecto estrogénico

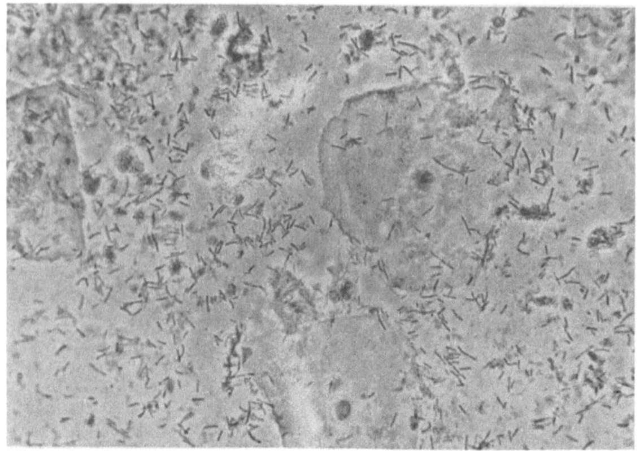

41

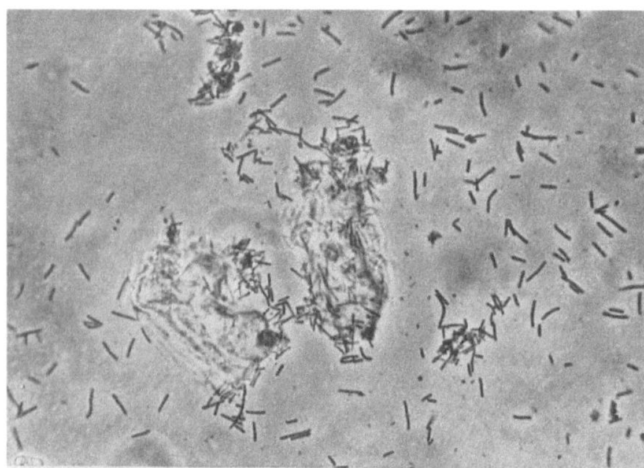

42

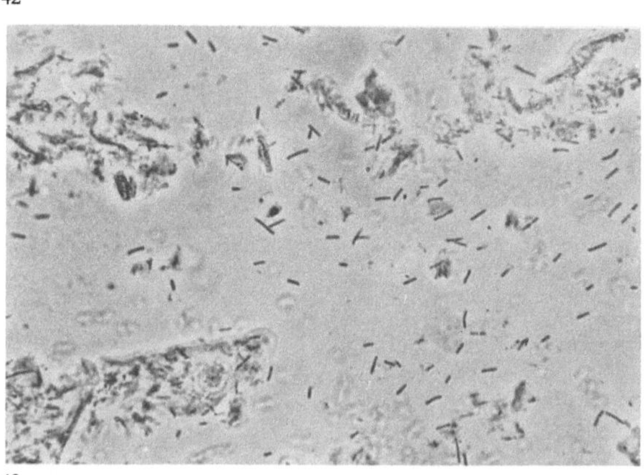

43

Vaginal Flora,
Döderlein's Bacillus

41. Smear of epithelial cells showing
estrogenic effect
Abundant Döderlein's bacilli,
rare polymorphonuclear leukocytes

42. Pure growth of Döderlein's bacilli

43. Pure growth of Döderlein's bacilli
Lysis of epithelial cells by Döderlein's
bacilli (Döderlein cytolysis).
Functional state of epithelial cells
obscured

Vaginalflora, Döderlein

41. Östrogener Ausstrich
Reichlich Döderleinkeime,
nur vereinzelte Leukocyten

42. Reine Döderleinflora

43. Reine Döderleinflora
Funktion nicht erkennbar, Döderleincytolyse

Flora vaginal, Döderlein

41. Frotis estrogénico
Rico en bacilos de Döderlein, sólo
escasos leucocitos

42. Flora de Döderlein no contaminada

43. Flora de Döderlein no contaminada
Función irreconocible, citolisis de
Döderlein

Vaginal Flora,
Döderlein's Bacillus

44. Flora of Döderlein's bacilli showing
lysis of epithelial cells
(Döderlein cytolysis). Only the nuclei
of intermediate cells remain.
The cytolysis is pronounced

45. Pregnancy
Chiefly intermediate cells. The microorganisms are mostly Döderlein's
bacilli with a few cocci. The cytolysis
is severe

46. Döderlein's bacilli among partially
lysed groups of intermediate cells
Functional state is obscured

Vaginalflora, Döderlein

44. Döderleinflora mit Döderleincytolyse
Übrig geblieben sind nur die Kerne
der Zellen. Es handelt sich um Kerne
von Intermediärzellen. Ausgeprägte
Döderleincytolyse

45. Gravidität
Vorwiegend Intermediärzellen, fast
ausschließlich Döderleinkeime,
Döderleincytolyse, daneben Kokken

46. Gruppen von Intermediärzellen,
dazwischen Döderleinkeime
Funktion schwer erkennbar

Flora vaginal, Döderlein

44. Flora y citolisis de Döderlein
Los núcleos son el único vestigio
celular que permanece. Se trata de
núcleos de células intermedias.
Pronunciada citolisis de Döderlein

45. Gestación
Preponderancia de células intermedias,
casi exclusivamente bacilos de
Döderlein, citolisis de Döderlein,
presencia de cocáceas

46. Grupos de células intermedias y
bacilos de Döderlein
Función difícil de reconocer

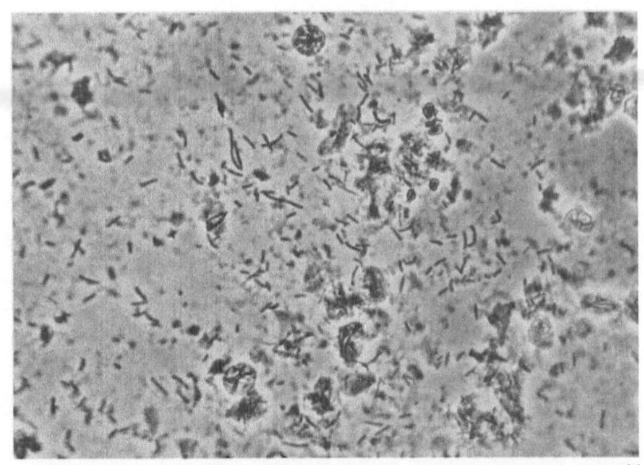

44

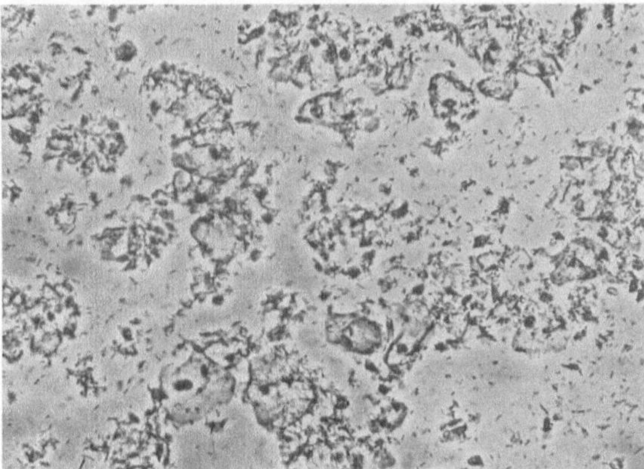

45

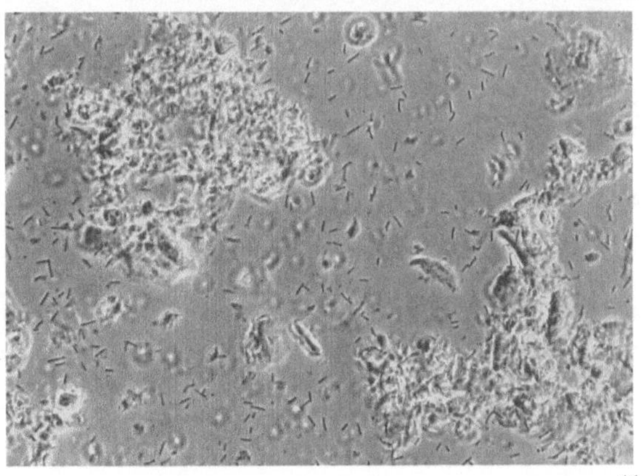

46

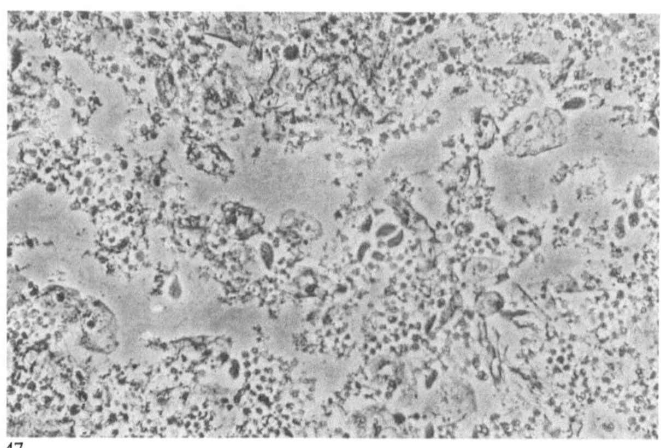

47

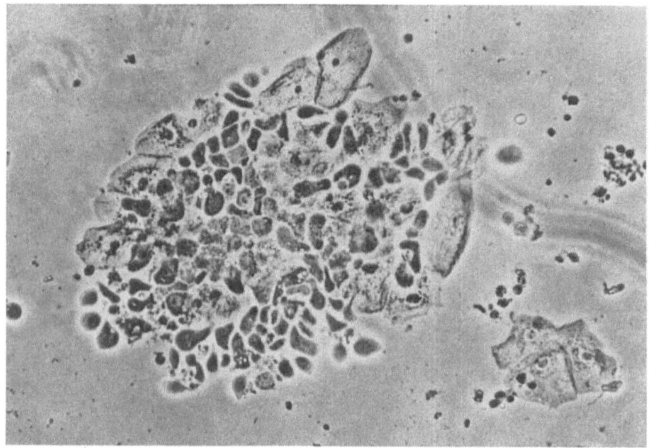

48

Microorganisms of the Vagina, Trichomonas	**Vaginalflora, Trichomonaden**	**Flora vaginal, Tricomonas**
47. Trichomonas colpitis Polymorphonuclear leukocytes, a few superficial and basal cells	47. Trichomonaden-Kolpitis Leukocyten, einige Superficial- und Basalzellen	47. Colpitis tricomoniásica Leucocitos, algunas células superficiales y basales
48. Groups of Trichomonads	48. Gruppe von Trichomonaden	48. Grupo de tricomonas

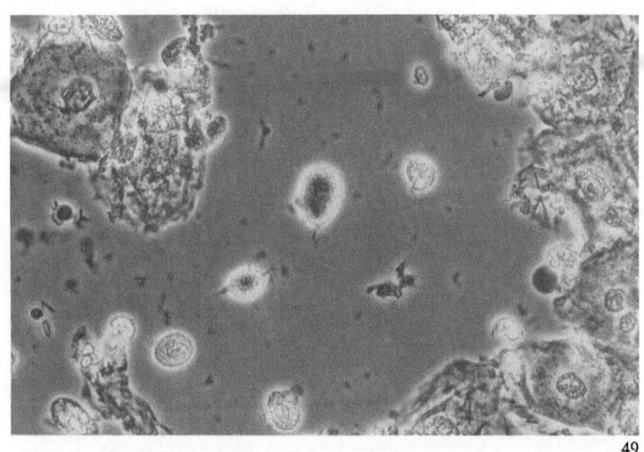

49

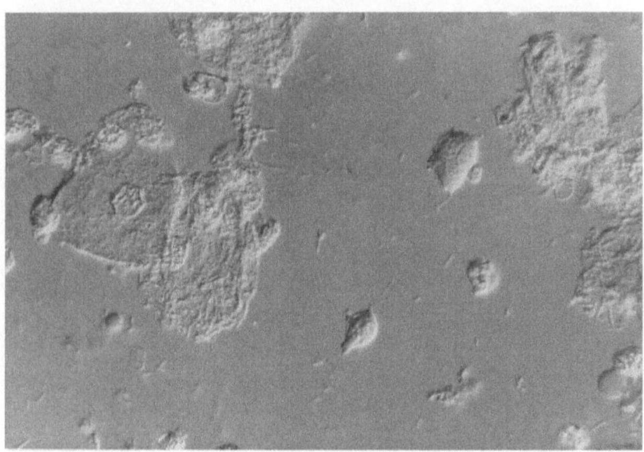

50

Microorganisms of the Vagina, Trichomonas	**Vaginalflora, Trichomonaden**	**Flora vaginal, Tricomonas**
49. and 50. Single Trichomonads (Phase-contrast and interference-contrast)	49. und 50. Einzelne Trichomonaden (Phasenkontrast und Interferenzkontrast)	49. y 50. Tricomonas aisladas (Contraste de fases e interferencia)

Microorganisms of the Vagina, Trichomonas

51. and 52. Single Trichomonad
with three flagella in front
and an undulating membrane
and trailing flagellum behind
(Phase-contrast and interference-contrast)

Vaginalflora, Trichomonaden

51. und 52. Einzelne Trichomonaden mit drei Geißeln
vorn und Schleppgeißel
(Phasenkontrast und Interferenzkontrast)

Flora vaginal, Tricomonas

51. y 52. Tricomonas aisladas con
flagelos y el punto de
implantación flagelar
(Contraste de fases e interferencia)

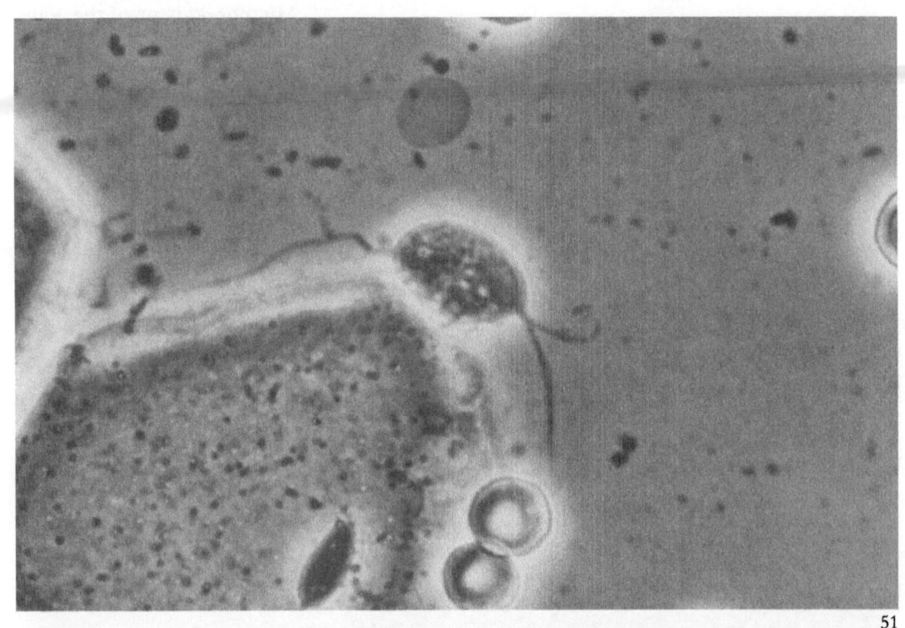

51

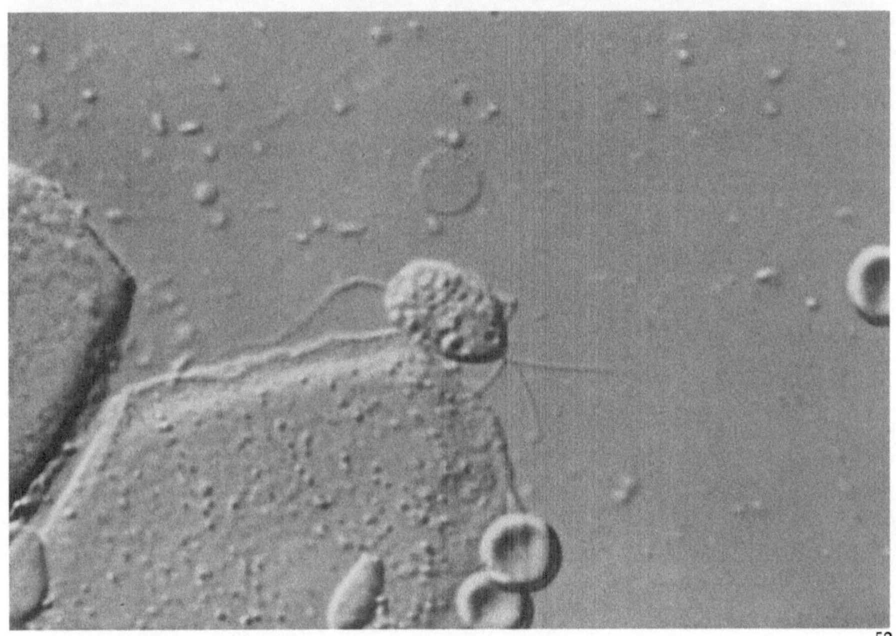

52

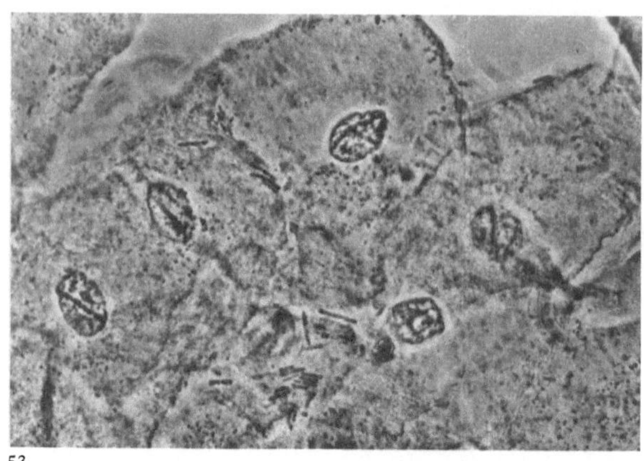

53

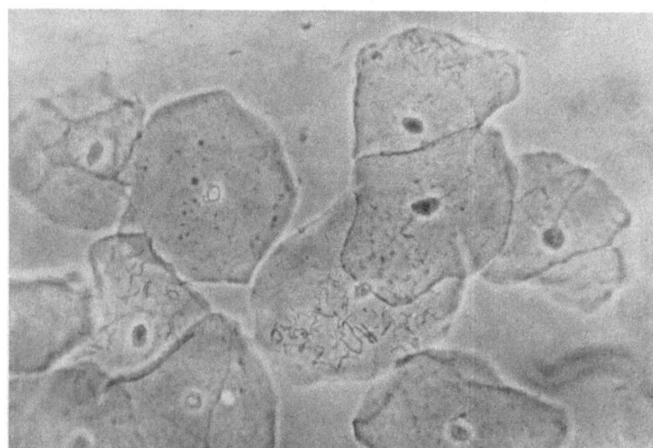

54

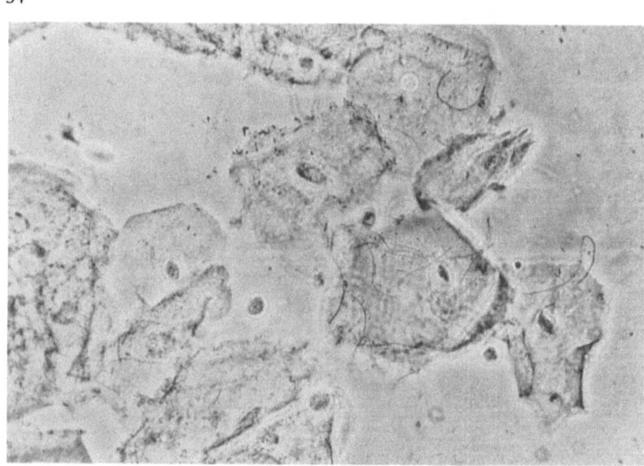

55

Microorganisms of the Vagina, Bacteria

53. Lactobacilli on intermediary cells
54.—56. Thread-like lactobacilli

Vaginalflora, Bakterien

53. Lactobacillen auf Intermediärzellen
54.—56. Fadenförmige Lactobacillen

Flora vaginal, Bacterianos

53. Lactobacilos en células intermedias
54.—56. Lactobacilos filiformes

Microorganisms of the Vagina, Bacteria

57. and 58. Lactobacilli from culture
(fixed and stained)

Vaginalflora, Bakterien

57. und 58. Lactobacillen aus der Kultur
(fixiert und gefärbt)

Flora vaginal, Bacterianos

57. y 58. Lactobacilos filiformes de un
cultivo (fijado y teñido)

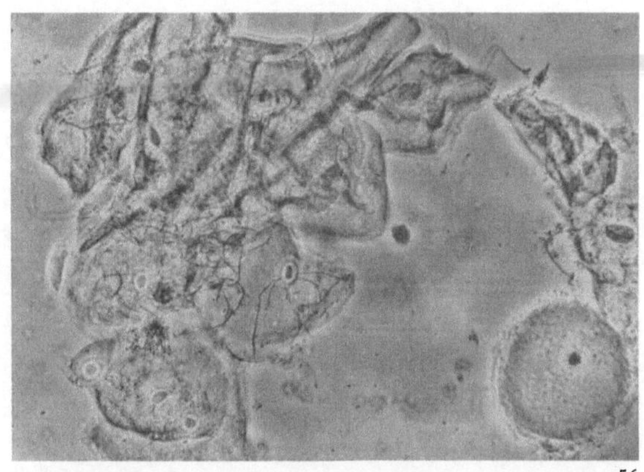

56

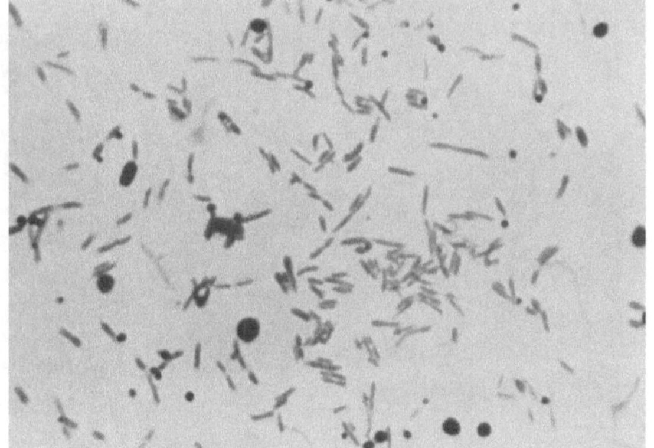

57

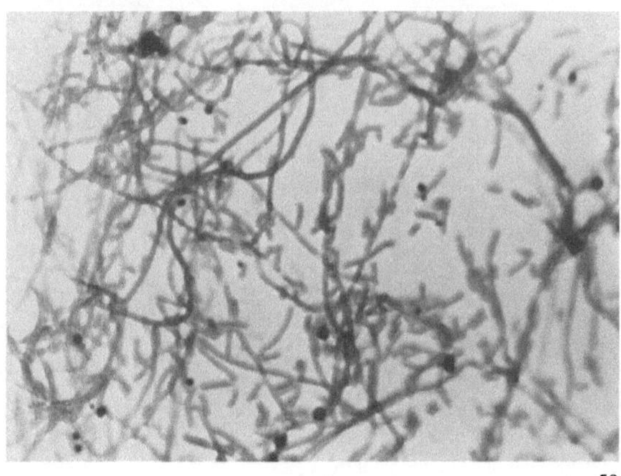

58

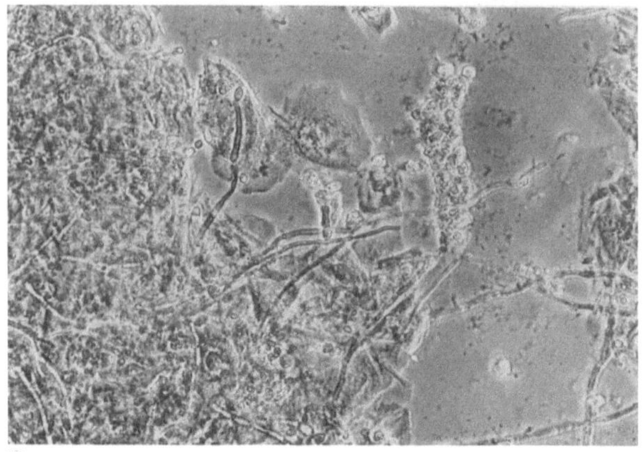

59

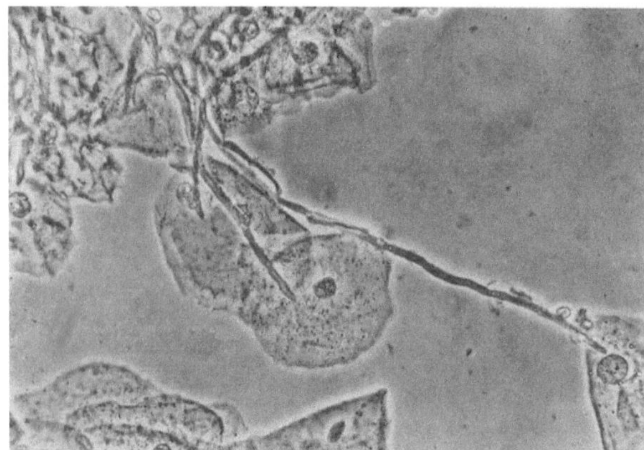

60

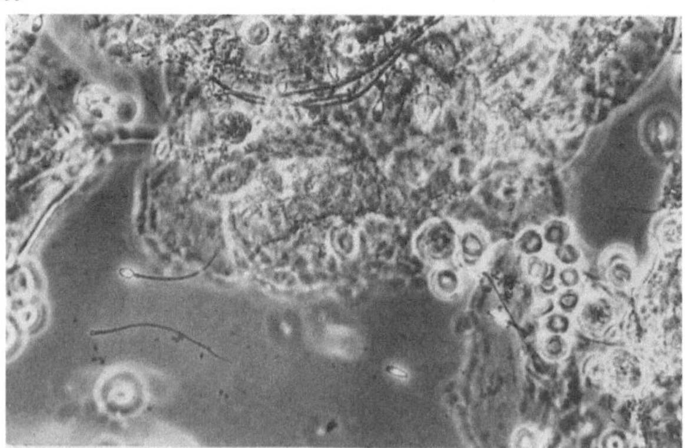

61

Vaginal Flora, Mycoses

59. Candida albicans
    Hyphae growing on clumps of
    epithelial cells, numerous poly-
    morphonuclear leukocytes

60. Formation of hyphae by Candida

61. Hemorrhagic colpitis
    Groups of erythrocytes; Candida
    albicans growing on epithelial cells;
    also sperm, one with a degenerated
    head

Vaginalflora, Mycosen

59. Monilia albicans
    Hyphenbildung, welche das Zellbild
    überlagert. Zahlreiche Leukocyten

60. Hyphenbildung von Candida

61. Hämorrhagische Kolpitis
    Gruppen von Erythrocyten. Die
    Epithel-Zellen werden von Candida
    albicans überlagert. Außerdem
    einige Spermien, davon eines mit
    degeneriertem Kopf

Flora vaginal, Micosis

59. Monilia
    Agrupación de hifas que recubren
    el cuadro celular. Abundantes
    leucocitos

60. Formación de hifas de Cándida
    albicans

61. Colpitis hemorrágica
    Grupo de eritrocitos, las células
    están recubiertas por Cándida
    albicans. Además algunos espermios,
    uno de ellos con extremidad
    cefálica en degeneración

Vaginal Flora, Mycoses

62. and 63. Hyphae of Candida albicans
Phase-contrast. Interference-
contrast

64. Hyphae of Candida albicans
growing on cells

Vaginalflora, Mycosen

62. und 63. Hyphen von Candida
albicans
Phasenkontrast und Interferenzkontrast

64. Candida albicans
Auflagerung von Hyphen auf
Zellen

Flora vaginal, Micosis

62. y 63. Hifas de Cándida albicans
Método de interferencia, contraste
de fases

64. Cándida albicans
Disposición de hifas sobre las
células

62

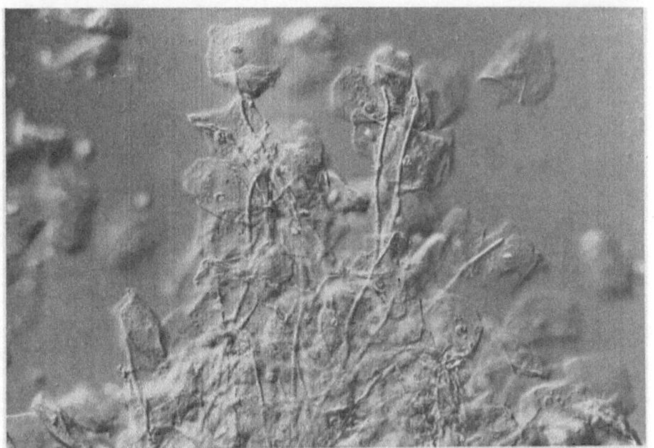

63

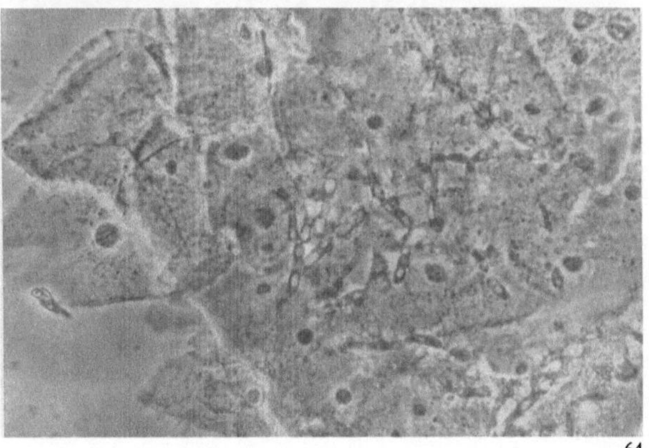

64

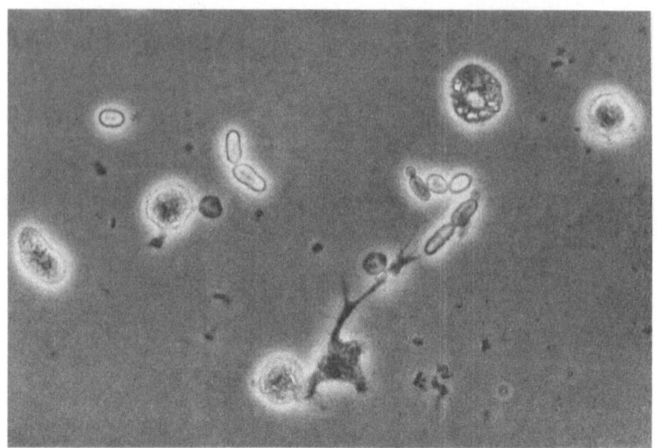

65

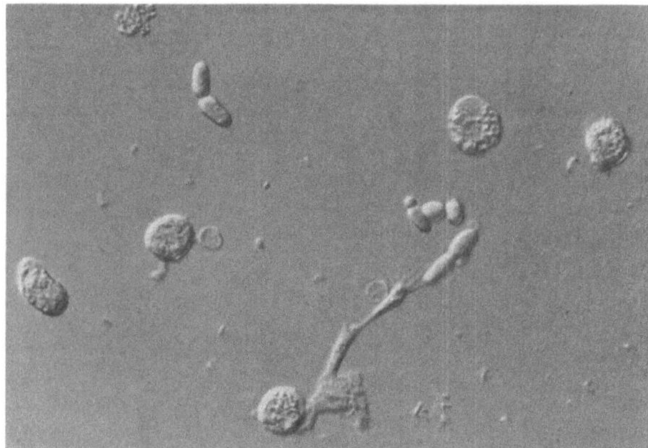

66

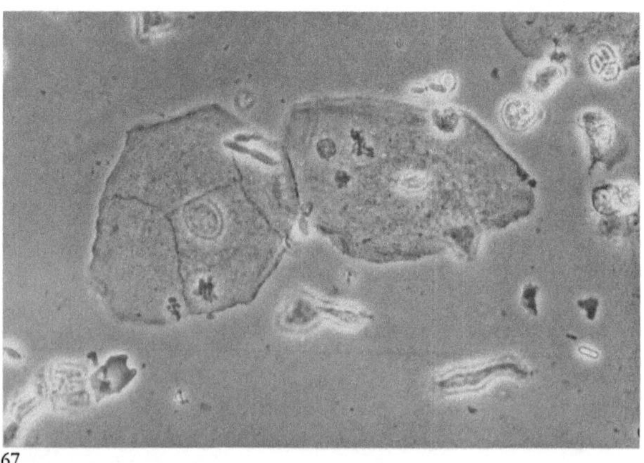

67

Vaginal Flora, Mycoses

65. and 66. Candida organisms
Budding of yeast-like spores

67. and 68. Candida organisms
Yeast-like spores

Vaginalflora, Mycosen

65. und 66. Candida albicans
Abschnürung von Sproßzellen

67. und 68. Candida albicans
Sproßzellen

Flora vaginal, Micosis

65. y 66. Cándida albicans
Hongos de forma fragmentada

67. y 68. Cándida albicans
Hongos de forma fragmentada

Vaginal Flora, Mycoses

69. and 70. Septated hyphae of Candida
organisms and long Döderlein's
bacilli

Vaginalflora, Mycosen

69. und 70. Sproßzellen und Fadenbakterien

Flora vaginal, Micosis

69. y 70. Hongos y bacterias filiformes

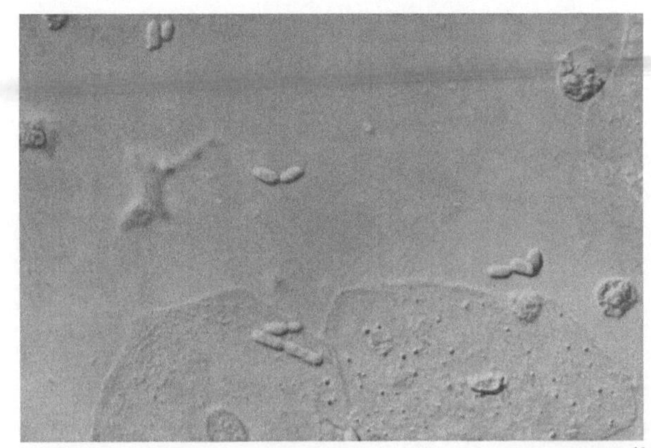

68

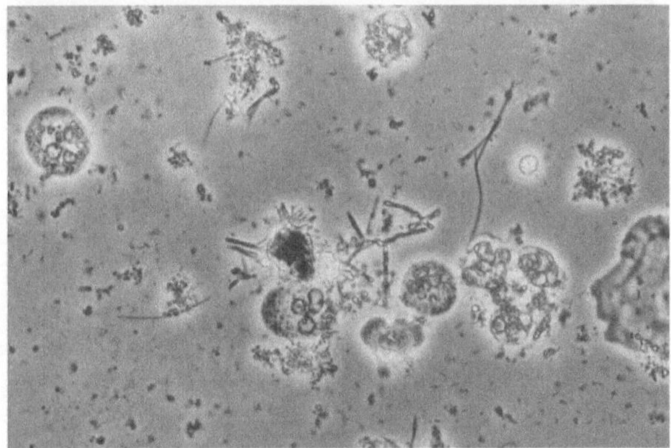

69

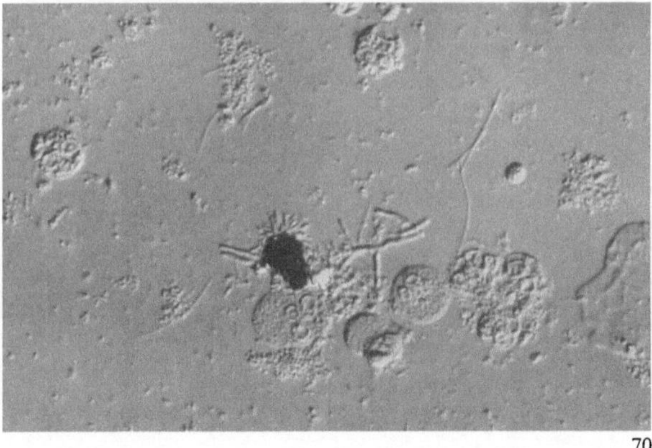

70

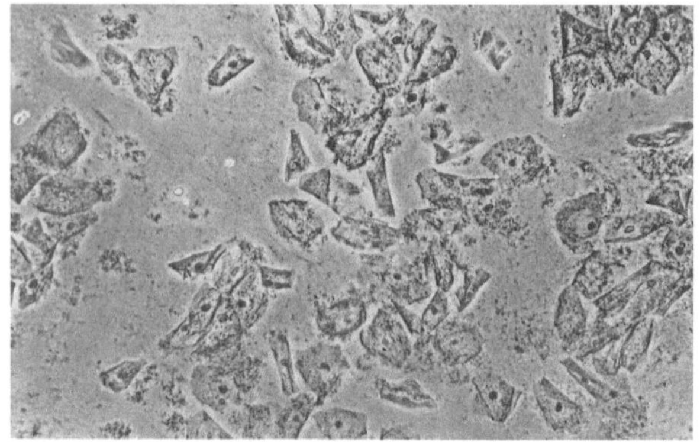

71

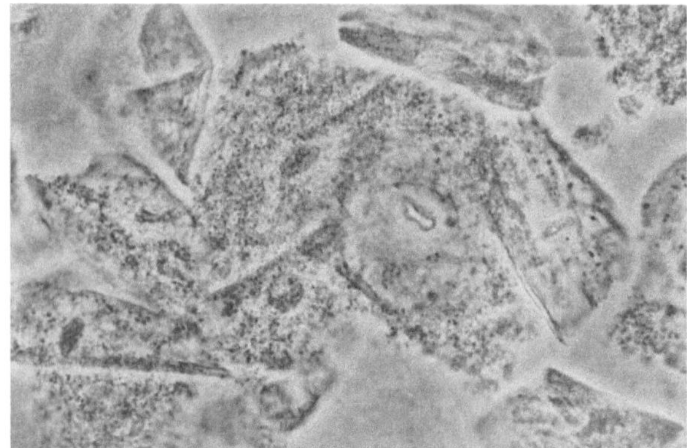

72

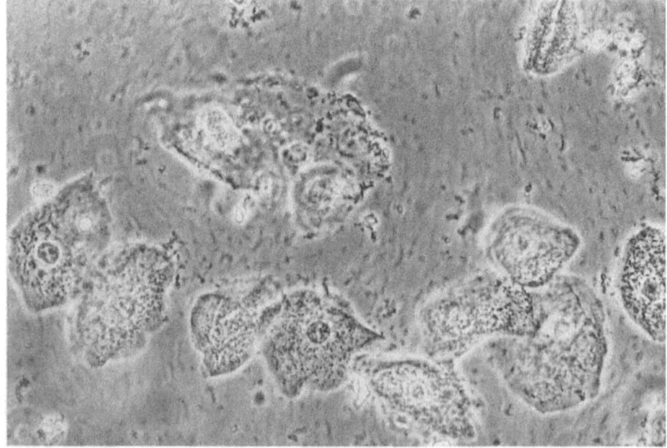

73

Vaginal Flora, Hemophilus Vaginalis

71. and 72. Colonies of small bacteria on the epithelial cells

73. Bacteria covegrin intermediary cells

Vaginalflora, Hämophilus vaginalis

71. und 72. Auflagerung auf Zellen, Bakterienrasen

73. Auflagerung auf Intermediärzellen

Flora vaginal, Hemófilus vaginalis

71. y 72. Disposición sobre las células y extendidos bacterianos

73. Disposición sobre células intermedias

Vaginal Flora, Hemophilus Vaginalis

74. Colonies of Hemophilus vaginalis and long bacilli

75. Aggregates of bacteria, some adherent to epithelial cells

76. Heavy growth of bacteria on epithelial cells

Vaginalflora, Hämophilus vaginalis

74. Bakterienrasen (Hämophilus vaginalis und Fadenbakterien)

75. Bakterienrasen und Auflagerung auf Zellen

76. Bakterienrasen und Auflagerung auf Zellen

Flora vaginal, Hemófilus vaginalis

74. Extendidos bacterianos (Hemófilus vaginalis y bacterias filiformes)

75. Extendidos bacterianos y disposición sobre las células

76. Extendidos bacterianos y disposición sobre las células

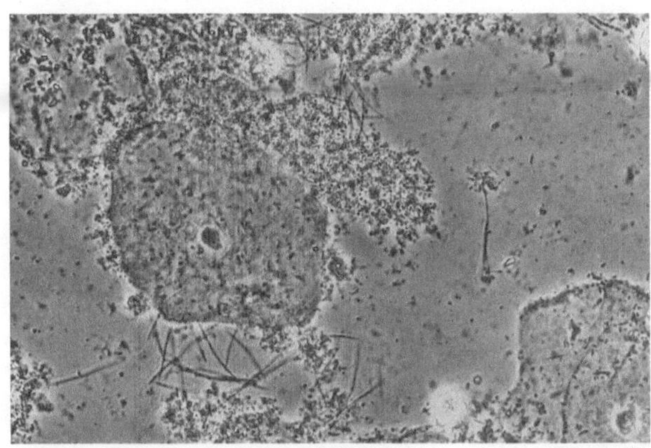

74

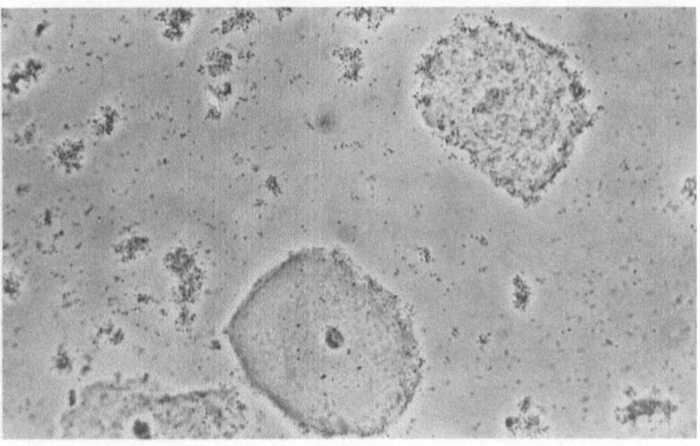

75

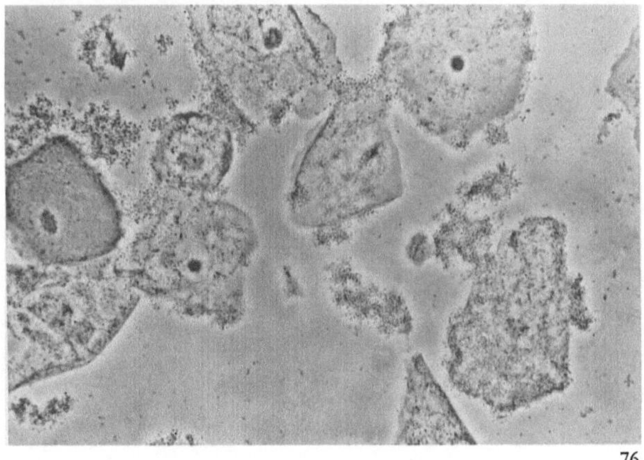

76

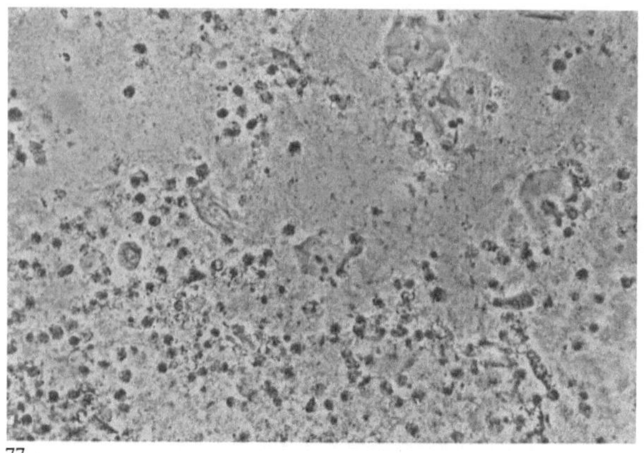

77

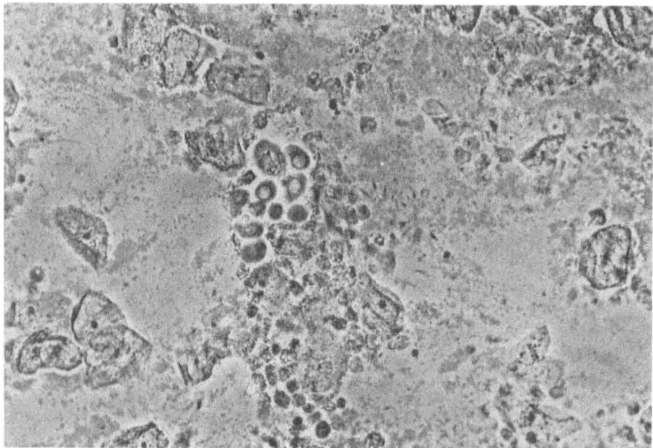

78

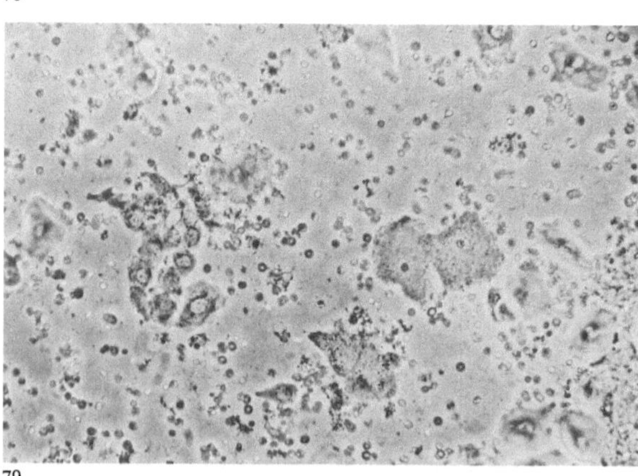

79

Vaginal Flora, Cocci

77. Myriads of cocci, some superficial cells, a few parabasal cells and lysis of cells by bacterial action

78. Flora of cocci, some lysis of cells by bacterial action, dyskaryosis due to the inflammation

79. Flora of cocci, erythrocytes, leukocytes, dyskaryosis

Vaginalflora, Kokken

77. Massenhafte Kokken, einige Superficialzellen, einige Parabasalzellen, stellenweise bakterielle Autolyse

78. Kokkenflora, z.T. Zellautolyse, einige entzündlich bedingte Dyskaryosen

79. Kokkenflora, Erythrocyten, Leukocyten, Dyskaryosen

Flora vaginal, Cocáceas

77. Cocáceas en masiva cantidad. Algunas células superficiales y parabasales. Citolisis bacteriana en algunas partes

78. Flora de cocáceas. En partes una autolisis celular. Algunas discariosis determinadas por la inflamación

79. Cocos, eritrocitos, leucocitos, discariosis celulares

Vaginal Flora, Cocci

80. and 81. Smear of atrophic
epithelium, flora of cocci.
(Phase-contrast and interference-
contrast)

82. Endocervicitis due to cocci, a few
ciliated columnar epithelial cells

Vaginalflora, Kokken

80. und 81. Ausstrich bei Atrophie,
Kokkenflora (Phasenkontrast und
Interferenzkontrast)

82. Endocervicitis mit Kokkenflora,
einige Endocervicalzellen mit Cilien

Flora vaginal, Cocáceas

80. y 81. Frotis atrófico, cocáceas
(contraste de fase e interferencia)

82. Endocervicitis con flora de
cocáceas, algunas células endo-
cervicales ciliadas

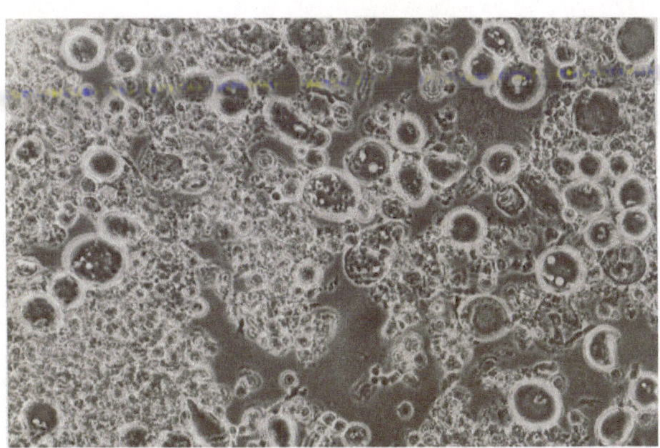

80

81

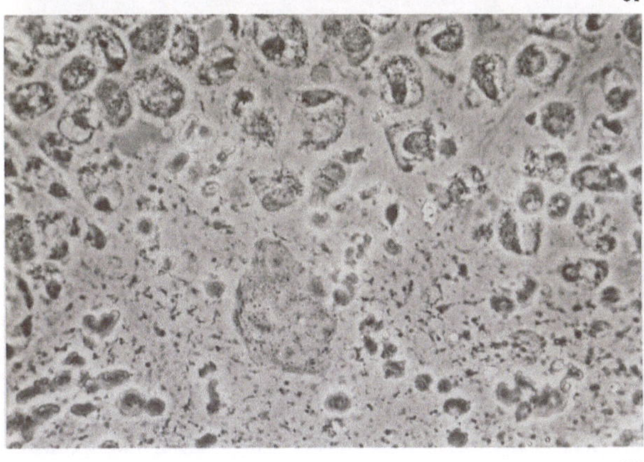

82

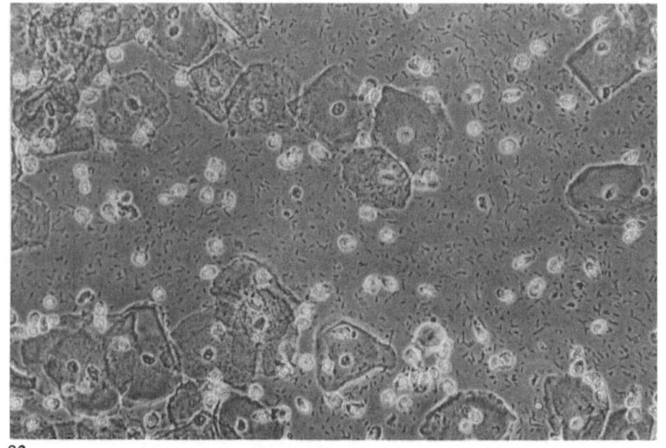

83

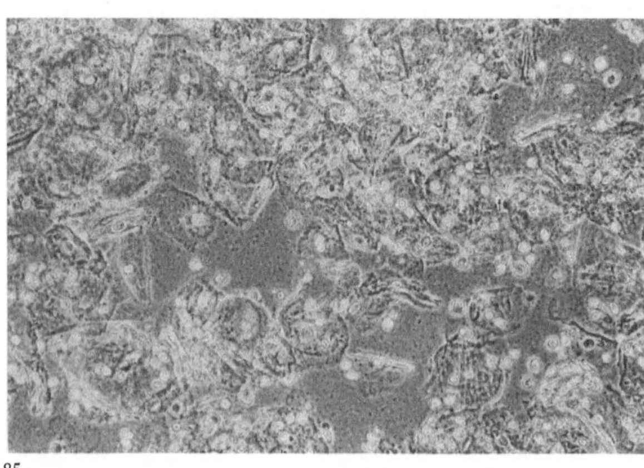

84

85

Vaginal Flora during Pregnancy

83. Fifth month of pregnancy
Flora of Döderlein's bacilli, many polymorphonuclear leukocytes. (Phase-contrast)

84. Sixth month of pregnancy
Flora of Döderlein's bacilli and rare cocci. (Phase-contrast)

85. Fourth month of pregnancy
Obvious mixed flora, cytological diagnosis is not possible. (Phase-contrast)

Vaginalflora bei Gravidität

83. Gravidität mens V
Döderleinflora, mäßig viele Leukocyten. (Phasenkontrast)

84. Gravidität mens VI
Döderleinflora und einzelne Kokken. (Phasenkontrast)

85. Gravidität mens IV
Ausgeprägte Mischflora. Diagnose cytologisch nicht möglich. (Phasenkontrast)

Flora vaginal y Gestación

83. 5° mes de gestación
Flora de Döderlein, más que mediana cantidad de leucocitos (fases contrastadas)

84. 6° mes de gestación
Flora de Döderlein y cocáceas escasas (fases contrastadas)

85. 4° mes de gestación
Presencia abundante de flora mixta. Diagnóstico citológico no es posible (fases contrastadas)

Vaginal Flora during Pregnancy

86. Fifth month of pregnancy
    Flora of Döderlein's bacilli, many
    polymorphonuclear leukocytes.
    (Interference-contrast)

87. Sixth month of pregnancy
    Flora of Döderlein's bacilli and
    rare cocci. (Interference-contrast)

88. Fourth month of pregnancy
    Obvious mixed flora, cytological
    diagnosis is not possible.
    (Interference-contrast)

Vaginalflora bei Gravidität

86. Gravidität mens V
    Döderleinflora, mäßig viele
    Leukocyten. (Interferenzkontrast)

87. Gravidität mens VI
    Döderleinflora und einzelne
    Kokken. (Interferenzkontrast)

88. Gravidität mens IV
    Ausgeprägte Mischflora. Diagnose
    cytologisch nicht möglich
    (Interferenzkontrast)

Flora vaginal y Gestación

86. 5° mes de gestación
    Flora de Döderlein, más que
    mediana cantidad de leucocitos
    (contraste de inteferencia)

87. 6° mes de gestación
    Flora de Döderlein y cocáceas
    (contraste de inteferencia)

88. 4° mes de gestación
    Presencia abundante de flora mixta.
    Diagnóstico citológico no es
    posible (contraste de inteferencia)

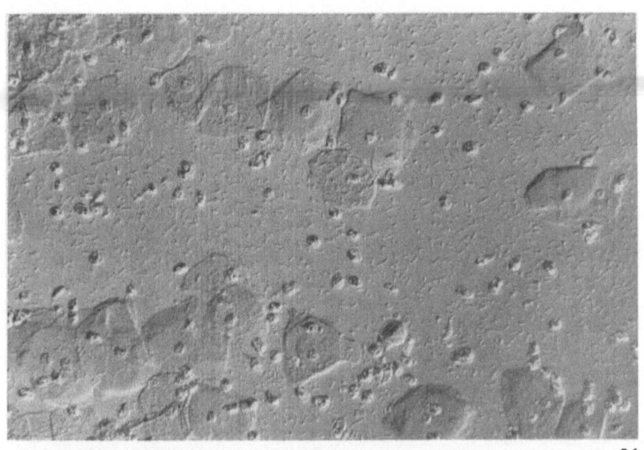

86

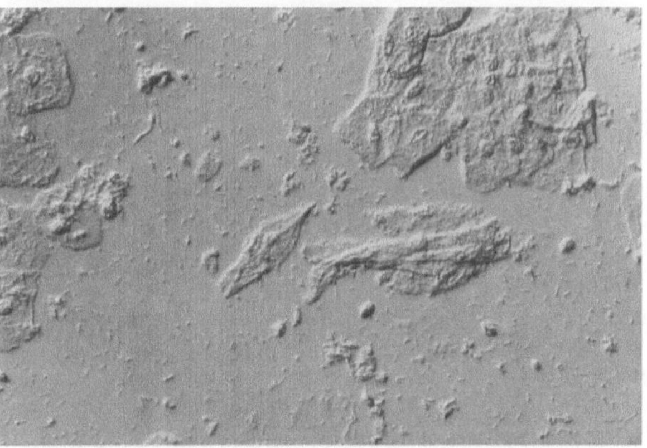

87

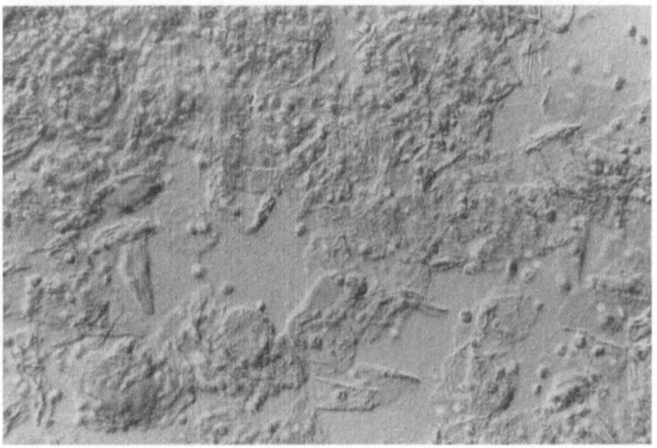

88

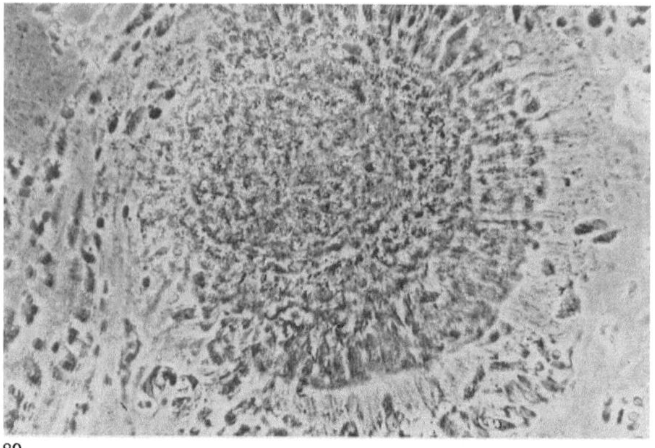

89

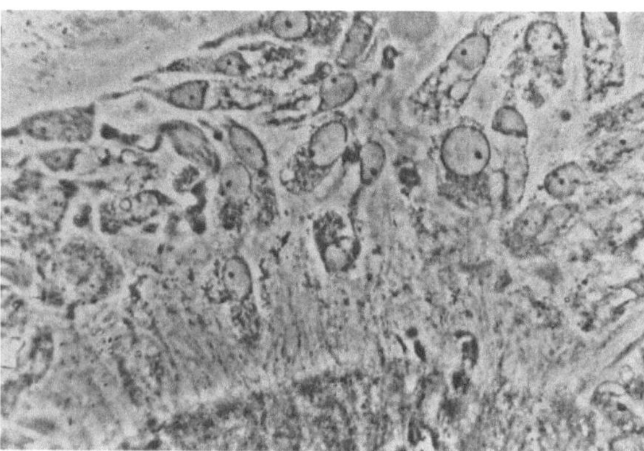

90

Endocervical Canal, Columnar Epithelium

89. Portion of an endocervical gland

The columnar epithelium, resting on a basement membrane, forms a ring of pallisading cells about clumped polymorphonuclear leukocytes and cellular debris. At the margin some columnar cells are being shed. Polymorphonuclear leukocytes in the vicinity. Diagnosis: endocervicitis

90. Enlargment of 89

Margin of an endocervical gland. The columnar epithelial cells have been shed from their associations. Pallisading present. The basement membrane is clearly evident. Many of the cells contain vacuoles of mucus. The ciliated border, although indistinct, reveals a regular rhythmic beating during study under the microscope

Cervicalkanal, Cylinder-epithel

89. Teil einer Cervixdrüse

Das Cylinderepithel sitzt der Basalmembran palisadenförmig auf. Im Bindegewebe zahlreiche Leukocyten. Am Rande schilfern die Zylinderepithelien ab. Auch in der Umgebung Leukocyten. Diagnose: Endocervicitis

90. Vergrößerung von 89

Rand einer Cervixdrüse. Die Cylinderepithelzellen lösen sich aus ihrem Verband. Noch deutliche Palisadenstellung. Haftplatte deutlich ausgebildet. Im Cytoplasma der Zellen Schleimvacuolen. Der Ciliensaum ist unscharf. Er zeigt bei mikroskopischer Beobachtung einen gleichförmigen Bewegungsrhythmus

Canal cervical, Epitelio cilíndrico

89. Porción de una glándula cervical

El epitelio cilíndrico se asienta sobre el tejido conjuntivo en forma de empalizada. En el tejido conjuntivo abundante infiltración leucocitaria. En los bordes, el epitelio cilíndrico se ha descamado. También en los alrededores hay leucocitos. Diagnóstico: endo-cervicitis

90. Mayor aumento de la fig. 89

Glándula cervical del borde. Las células del epitelio cilíndrico se desprenden de su implantación. Todavía evidente disposición en empalizada. En el citoplasma celular, vacuolas de secreción mucosa. El reborde ciliado es poco nítido. A la observación microscópica estos cilios muestran un movimiento rítmico de sentido uniforme

Endocervical Canal,
Columnar Epithelium

91. Exfoliated columnar cells lying singly
    The footplate and ciliated border
    are distinct, the nucleus is round,
    large, vesicular, and empty except
    for a nucleolus. During observation
    under the microscope the cilia beat
    rhythmically

92. Columnar cells from the vagina
    Desquamated clumps of columnar
    epithelial cells (from an ectopia), in
    part altered by inflammation. In the
    vicinity numerous naked nuclei;
    polymorphonuclear leukocytes and
    single erythrocytes

Cervicalkanal, Cylinder-
epithel

91. Abgeschilferte, einzeln liegende
    Cylinderepithelien
    Deutliche Fußplatte, deutlicher
    Bürstensaum. Der Kern ist rund,
    groß, bläschenförmig, leer, bis auf
    einen Nucleolus; unter dem
    Mikroskop deutliche rhythmische
    Bewegungen der Cilien

92. Entnahme aus dem Vaginalraum
    Aus einer Ektopie abgeschilferter
    Cylinderepithelkomplex, entzünd-
    lich verändert. In der Umgebung
    zahlreiche nackte Kerne, Leuko-
    cyten, einzelne Erythrocyten

Canal cervical,
Epitelio cilíndrico

91. Epitelio cilíndrico exfoliado y
    aislado
    Nítida zona de implantación y
    escobilla ciliar. El núcleo es
    redondeado, grande, vesicular,
    vacío o bien con un nucléolo. Bajo
    el microscopio hay un nítido
    movimiento rítmico de los cilios

92. Muestra tomada de la vagina
    Epitelio cilíndrico con cambios
    inflamatorios, exfoliado de una
    ectopía cervical. En las cercanías
    abundantes núcleos desnudos,
    leucocitos, algunos eritrocitos

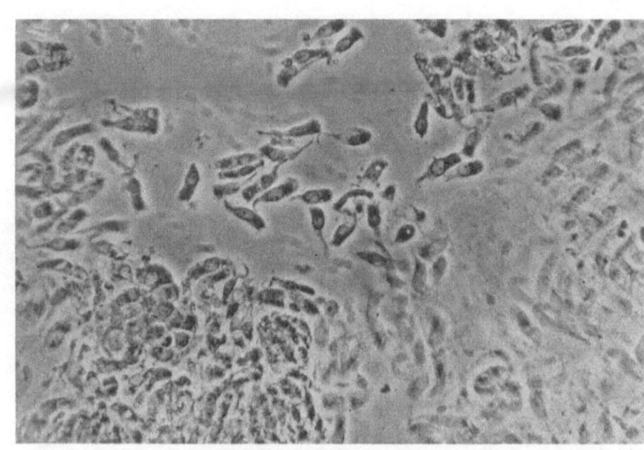

91

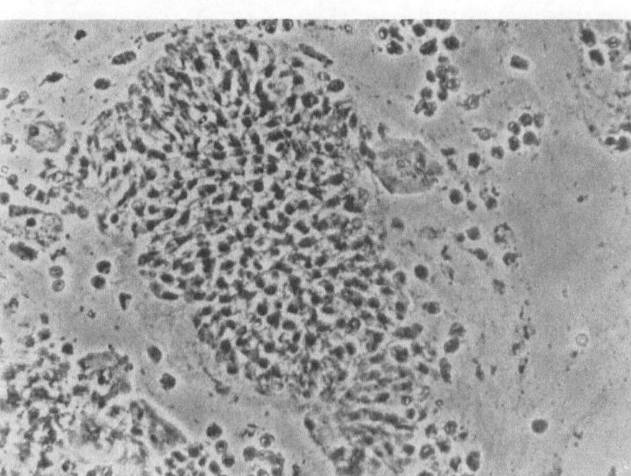

92

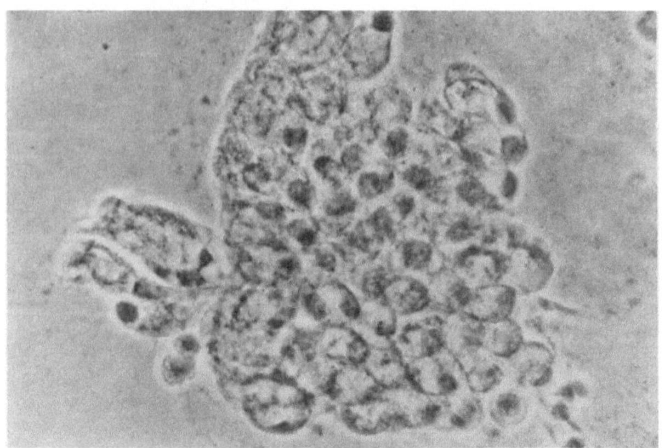

93

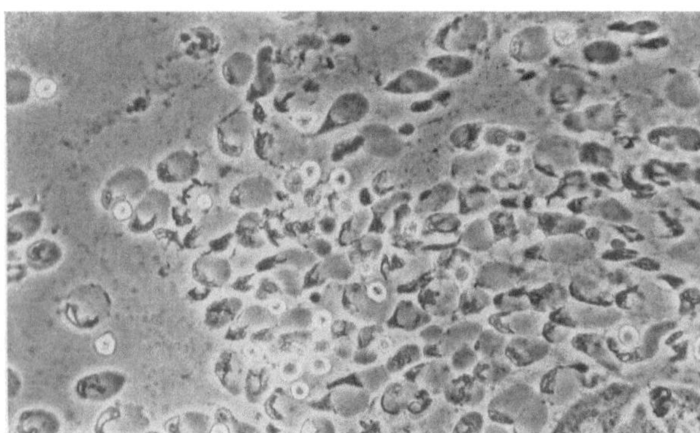

94

Endocervical Canal, Columnar Epithelium

93. An aggregate of degenerating, columnar epithelial cells

    Typical honey-combed appearance, cytoplasm in part disintegrated. Nuclei distinctly swollen

94. Advanced degeneration of columnar epithelial cells

    The ciliated border has been shed. The cytoplasm up to the basal plate has disintegrated, distinct edema of the nuclei, numerous polymorphonuclear leukocytes nearby

Cervicalkanal, Cylinder-epithel

93. Cylinderepithelkomplex in Degeneration

    Typische Honigwabenstruktur, Cytoplasma zum Teil aufgelöst, deutliche Quellung der Kerne

94. Fortgeschrittene Degeneration der Cylinderepithelien

    Der Ciliensaum ist abgestoßen, das Protoplasma bis auf die Basalplatte aufgelöst, deutliches Kernödem, daneben zahlreiche Leukocyten

Canal cervical, Epitelio cilíndrico

93. Complejo de epitelio cilíndrico en degeneración

    Típica estructura en panal de abejas. Citoplasma en parte disuelto, salida de los núcleos

94. Avanzada degeneración del epitelio cilíndrico

    El reborde ciliar se ha desprendido. El protoplasma se halla disuelto hasta su porción basal, edema nuclear evidente. En las cercanías abundantes leucocitos

Endocervical Canal, Columnar Epithelium, Secondary Changes

95. Aggregate of endocervical epithelial cells in endocervicitis

    The fragment of epithelium is infiltrated with polymorphonuclear leukocytes and erythrocytes

96. Two columnar cells showing secondary changes

    Nuclear edema, formation of vacuoles in the cytoplasm, loss of ciliated border

Cervicalkanal, Cylinderepithel, sekundäre Veränderungen

95. Zellkomplex aus dem Cervicalkanal bei Endocervicitis

    Das Gewebsstück ist durchsetzt von Leukocyten und Erythrocyten

96. Zwei Cylinderepithelien mit sekundären Veränderungen

    Kernödem, Vacuolenbildung im Cytoplasma. Verlust des Ciliensaumes

Canal cervical, Transformaciones secundarias del epitelio cilíndrico

95. Complejo celular del canal cervical en una endocervicitis

    El trozo de tejido está infiltrado por leucocitos y eritrocitos

96. Dos células epiteliales cilíndricas con transformaciones secundarias

    Edema nuclear, formación de vacuolas en el citoplasma. Pérdida de la cubierta ciliada

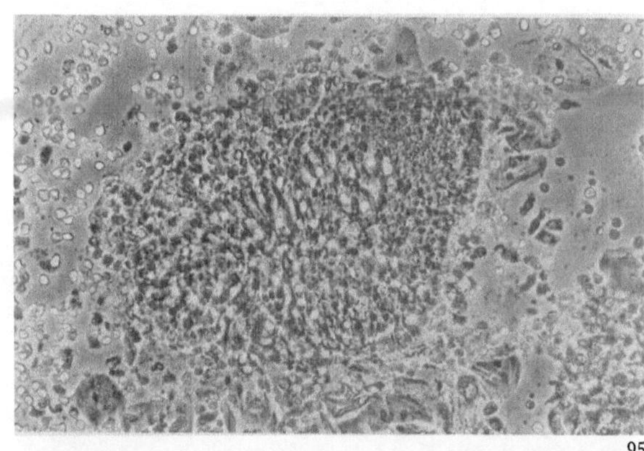

95

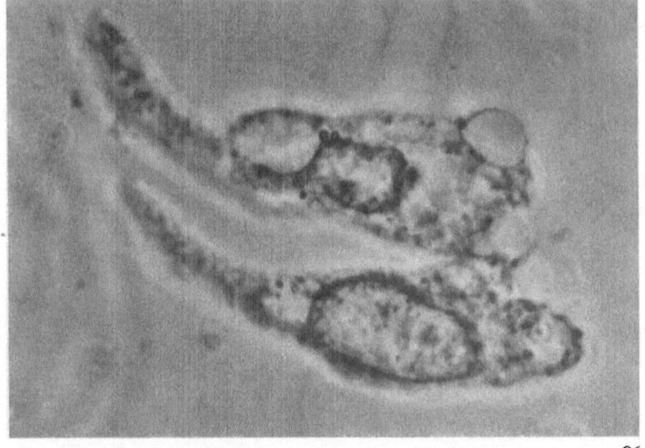

96

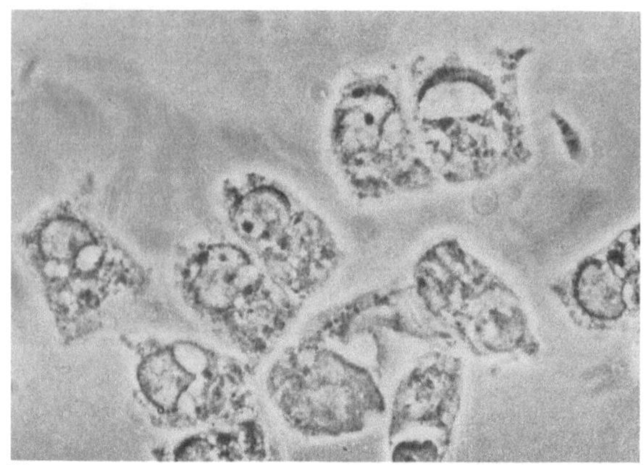

97

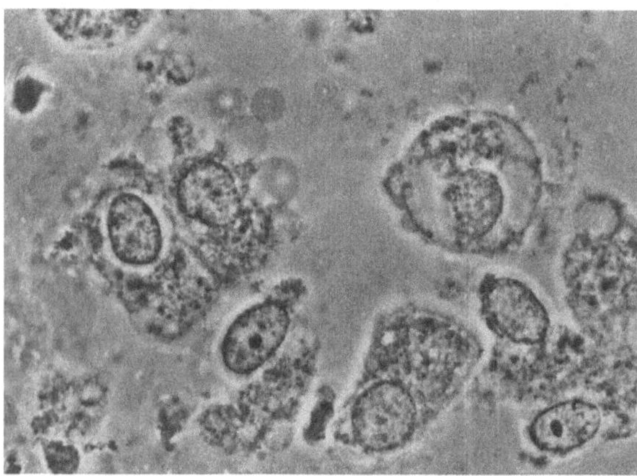

98

Endocervical Canal, Columnar Epithelium, Secondary Changes

97. Epithelial cells lying singly, showing secondary changes

    Formation of vacuoles in the cytoplasm, loss of ciliated border, formation of vacuoles in the nucleus, indistinct cell margins

98. Epithelial cells lying singly, showing intense secondary changes

    Nuclear swelling, disintegration of the cytoplasm; these cells may be mistaken for malignant cells

Cervicalkanal, Cylinderepithel, sekundäre Veränderungen

97. Einzeln liegende Cylinderepithelzellen, sekundäre Veränderungen

    Vacuolenbildung im Cytoplasma, Verlust des Bürstensaumes. Vacuolenbildung im Kern. Unschärfe der Zellgrenzen

98. Einzeln liegende Cylinderepithelzellen, starke sekundäre Veränderungen

    Kernquellung, Auflösung des Cytoplasmas. Verwechslung mit malignen Zellen möglich

Canal cervical, Transformaciones secundarias del epitelio cilíndrico

97. Algunas células ciliadas de epitelio cilíndrico con transformaciones secundarias

    Formación de vacuolas en el citoplasma. Pérdida de la cubierta ciliar. Vacuolas nucleares. Límites celulares no bien definidos

98. Células de epitelio cilíndrico con intensas transformaciones secundarias

    Imbibición nuclear. Disolución del citoplasma. Posible confusión con células malignas

Endcervical Canal,
Columnar Epithelium,
Secondary Changes

99. A group of columnar epithelial cells
in vaginal secretions
The columnar epithelial cells have
shrunken; only a narrow rim of
cytoplasm remains. The anchoring
basal corpuscle is distinct. Also in
the photograph are intermediary
cells which show dyskaryosis

100. Columnar epithelial cells in vaginal
secretions
Shrinkage of the cytoplasm with
extrusion of fluid, formation of
vesicles

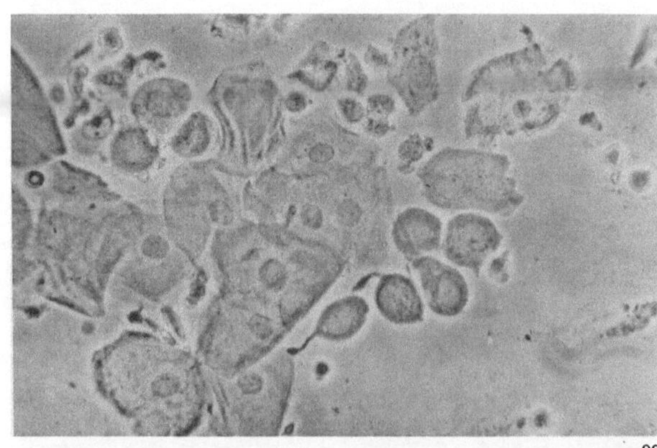

99

Cervicalkanal,
Cylinderepithel,
sekundäre Veränderungen

99. Eine Gruppe von Cylinderepithelzellen im Vaginalsekret
Die Zellen sind geschrumpft, es
ist nur ein schmaler Cytoplasma-
saum erhalten. Die basale
Verankerung ist deutlich. Im Bild
außerdem Intermediärzelle mit
Dyskariose

100. Cylinderepithelzellen im Vaginal-
sekret
Schrumpfung des Cytoplasmas mit
Flüssigkeitsaustritt. Blasenbildung

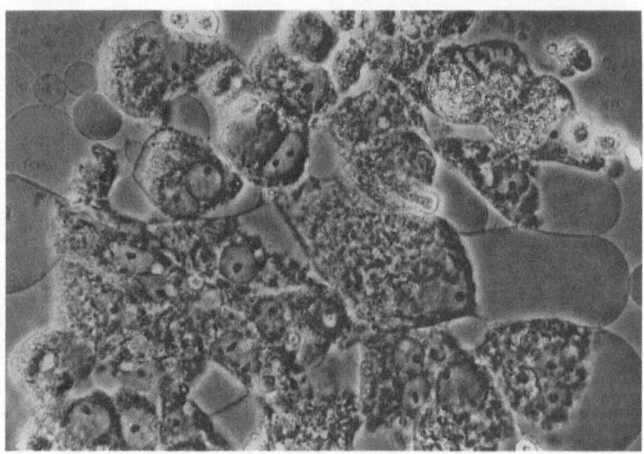

100

Canal cervical,
Transformaciones secundarias
del epitelio cilíndrico

99. Un grupo de células epiteliales
cilíndricas en la secreción vaginal
Las células están retraídas. Sólo
conservan una delgada cubierta
protoplasmática. La porción de
implantación basal es nítida. En
el cuadro además, células intermedias con discariosis

100. Células cilíndricas en la secreción
vaginal
Retracción del citoplasma con
pérdida de líquido. Formación de
vesículas

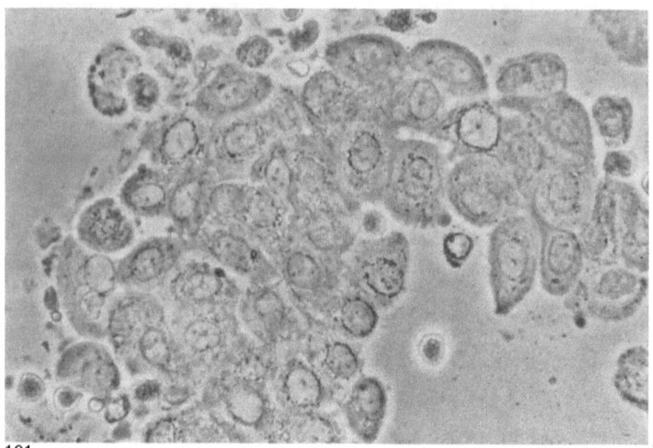

101

Ectropion, Squamous (Cell) Metaplasia

101. Poorly differentiated cells from a harmless squamous metaplasia of the endocervix
   The cells and nuclei regular in shape, the cytoplasm well-preserved

102. Cells from a harmless squamous metaplasia
   Secondary changes with disintergration of the cytoplasm

103. Well-preserved metaplastic cells with distinct cytoplasm; nearby cells with obscured cell boundaries

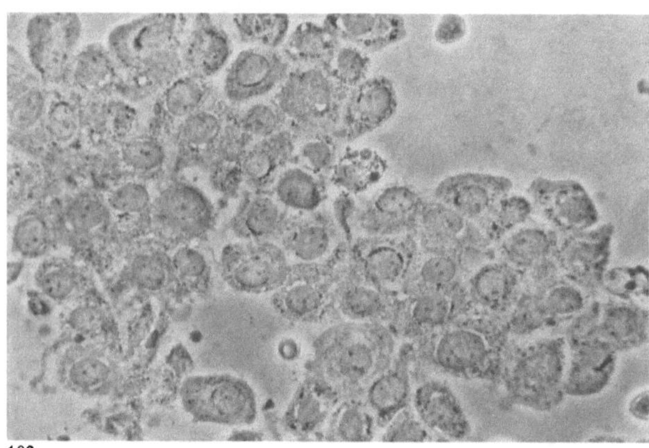

102

Ectropium, Plattenepithel-metaplasie

101. Wenig differenzierte Zellen aus einer gutartigen Plattenepithel-metaplasie im Cervicalkanal
   Gleichmäßige Kern- und Zellform, Plasma durchweg erhalten

102. Gutartige Zellen aus einer Platten-epithelmetaplasie
   Sekundäre Veränderungen durch Auflösung des Cytoplasmas

103. Guterhaltene metaplastische Zellen mit gutgeformtem Cytoplasma, daneben Zellen mit Verlust der Cytoplasmagrenzen

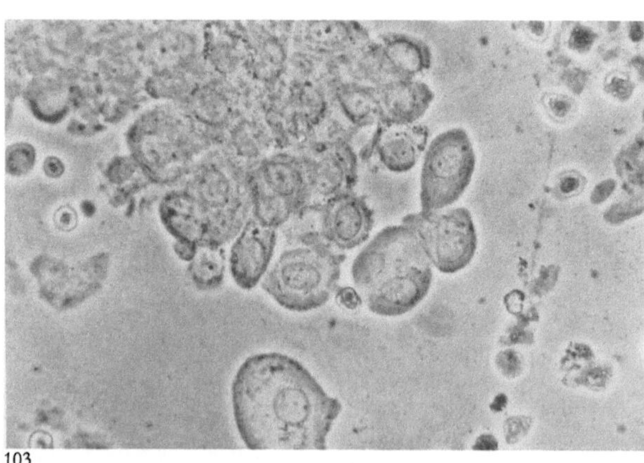

103

Ectropion, Metaplasia de epitelio epidermoídeo

101. Células poco diferenciadas de una metaplasia epidermoídea benigna en el canal cervical
   Regularidad en la forma de núcleos y células. Citoplasma en general conservado

102. Células benignas de una metaplasia epidermoídea benigna
   Transformaciones secundarias debido a una disolución del citoplasma

103. Células metaplásticas con un bien conformado citoplasma
   En las cercanías células con pérdida de los límites cito-plasmáticos

Ectropion, Squamous (Cell) Metaplasia

104. Proliferating cells of the basal type from a region of squamous metaplasia

105. Metaplastic cells
Highly differentiated

106. Metaplastic cells
From a region of squamous metaplasia. Three forms of cells. Some cells have become polygonal; the nucleus is that of a metaplastic cell

Ectropium, Plattenepithelmetaplasie

104. Proliferierende Zellen vom Typ der Basalzelle aus dem Bereich einer Plattenepithelmetaplasie

105. Metaplastische Zellen
Höhere Differenzierung

106. Metaplastische Zellen
Aus dem Bereich einer Plattenepithelmetaplasie. Drei Zellformen. Einige Zellen nehmen polygonale Formen an; der Kern entspricht dem einer metaplastischen Zelle

Ectropion, Metaplasia de epitelio epidermoídeo

104. Células proliferadas de tipo basal, que provienen de una metaplasia epidermoídea

105. Células metaplásticas
Elevada diferenciación

106. Células metaplásticas
Originadas en una zona de metaplasia epidermoídea. Células de tres formas distintas. Algunas células de forma poligonal, los núcleos corresponden a una célula metaplásica

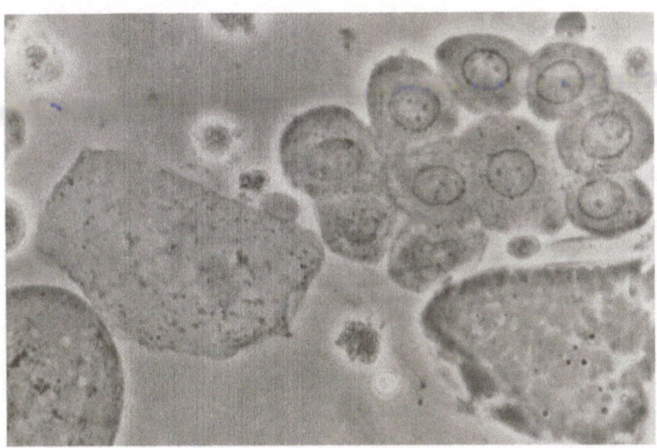

104

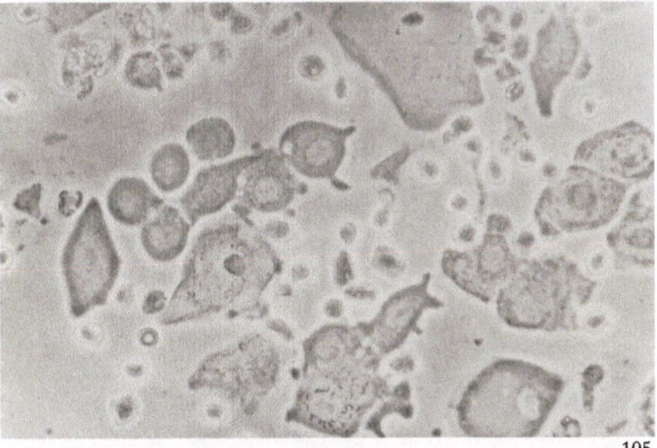

105

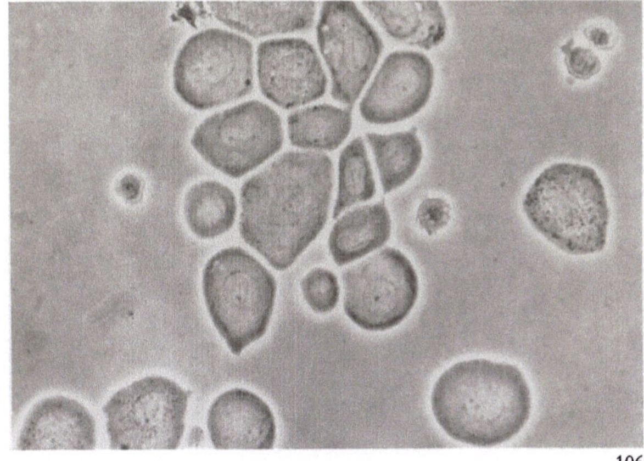

106

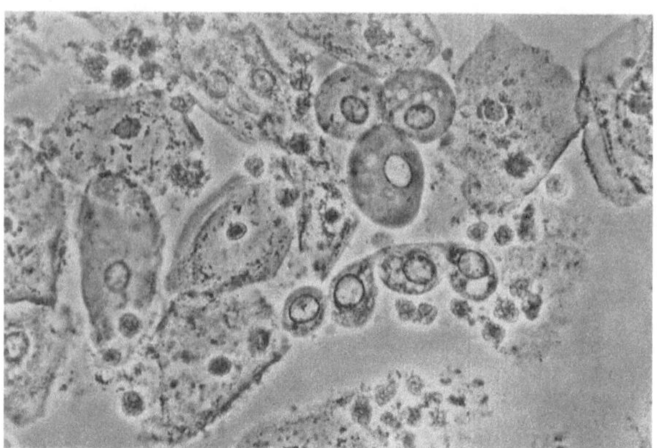

107

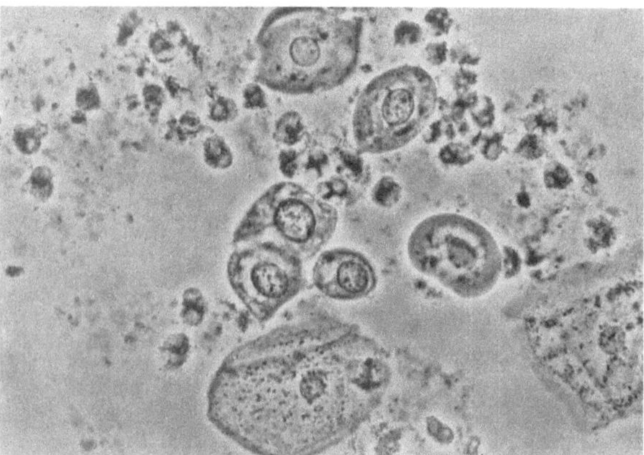

108

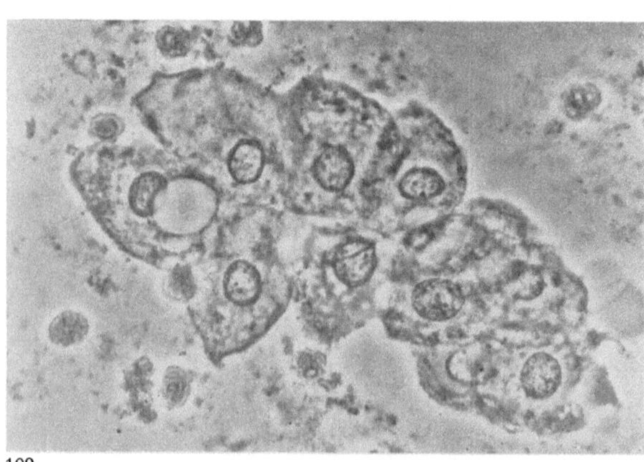

109

Dysplasia

107. Smear of a dysplastic ectocervical epithelium (portio vaginalis)

Intermediary cells and superficial cells with beginning pyknosis.
In the center a row of four basal cells, above these a group of three parabasal cells whose nuclear structure varies

108. Next to two normal intermediary cells a group of dyskaryotic parabasal cells. The nuclear structure and size variable

109. An aggregate of dyskaryotic intermediary cells

Polymorphonuclear leukocytes scattered nearby

Dysplasie

107. Abstrich von einem dysplastischen Portioepithel

Intermediär- und Superficialzellen mit beginnender Kernpyknose.
Im Zentrum 4 regelrechte Basalzellen in einer Reihe. Darüber eine Gruppe von 3 Parabasalzellen mit unterschiedlicher Kernstruktur

108. Neben zwei regelrechten Intermediärzellen eine Gruppe von dyskaryotischen Parabasalzellen. Kerngröße und Kernstruktur sind unterschiedlich

109. Intermediärzellgruppe — Dyskaryose

In der Umgebung Leukocyten

Displasia

107. Frotis de un epitelio exocervical displásico

Células intermedias y superficiales con picnosis nuclear inicial. En el centro 4 células basales en una línea. Algo más arriba un grupo de 3 células parabasales con diversa estructura nuclear

108. Junto a dos células intermedias normales, un grupo de células parabasales discarióticas. Se pueden apreciar diferencias en el tamaño y estructura de los núcleos

109. Grupo de células intermedias con discariosis

En los alrededores hay leucocitos

Dysplasia

110. Dyskaryosis of intermediary cells
     The nuclei resemble those of basal
     cells. The cytoplasm, however, is
     typically that of intermediary cells

111. Dyskaryosis of superficial cells

112. Dyskaryosis of intermediary and
     superficial cells

Dysplasie

110. Intermediärzelldyskaryose
     Die Kerne haben den Charakter von
     Basalzellen. Das Cytoplasma entspricht
     dem der Intermediärzellen

111. Superficialzellen mit Dyskaryose

112. Intermediär- und Superficialzellen
     mit Dyskaryose

Displasia

110. Discariosis de células intermedias
     Los núcleos tienen el caracter de
     células basales, el citoplasma
     corresponde a células intermedias

111. Células superficiales con discariosis

112. Células intermedias y superficiales
     con discariosis

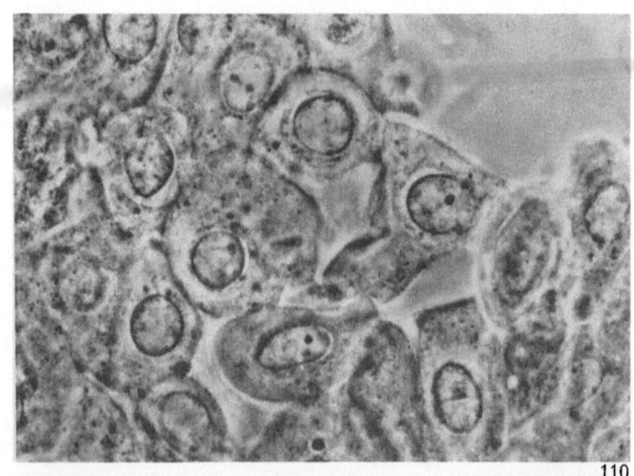

110

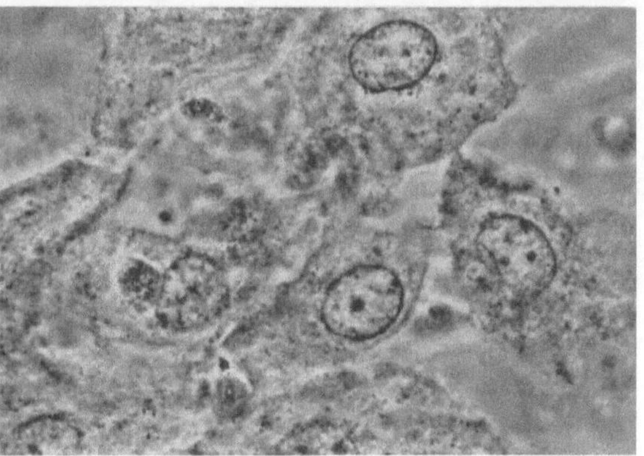

111

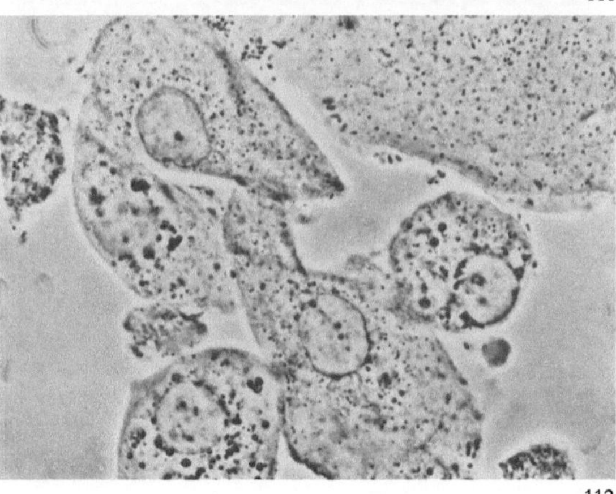

112

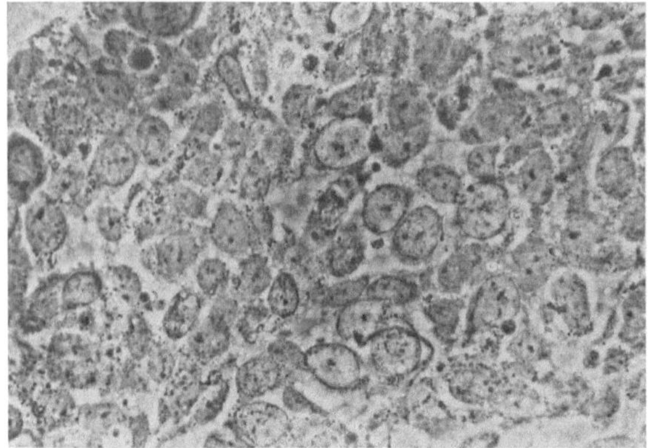

113

114

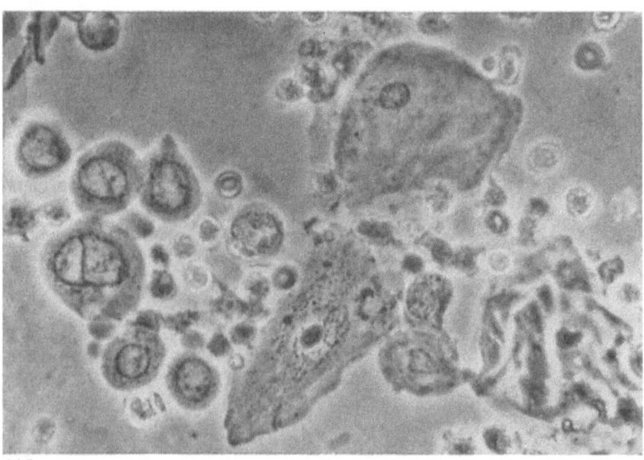

115

Carcinoma

113. Immature squamous cell carcinoma
The cytoplasm has disintegrated, the nuclei are irregular and small

114. Immature squamous cell carcinoma
The cytoplasm has partially disintegrated; there is little pleomorphism, the nuclei generally small

115. Moderately mature squamous cell carcinoma
Near two normal appearing superficial cells are several tumor cells one of which has a double nucleus. Their cytoplasm is well preserved. The tumor cells have the size of basal cells

Carcinom

113. Unreifes Plattenepithelcarcinom
Das Cytoplasma ist aufgelöst, die Kerne sind unregelmäßig, klein

114. Unreifes Plattenepithelcarcinom
Das Cytoplasma ist zum Teil aufgelöst, geringe Polymorphie der im ganzen kleinen Kerne

115. Mittelreifes Plattenepithelcarcinom
Neben zwei regelrechten Superficialzellen ein Tumorzellkomplex, darunter eine doppelkernige Zelle. Das Cytoplasma ist gut erhalten. Die Zellen entsprechen in ihrer Größe Basalzellen

Carcinoma

113. Carcinoma epidermoide inmaduro
El citoplasma está disuelto, el núcleo es irregular, pequeño

114. Carcinoma epidermoide inmaduro
El citoplasma está en parte disuelto, los núcleos en general pequeños presentan escasa polimorfía

115. Carcinoma epidermoide de mediana madurez
Junto a células superficiales normales, un complejo de células tumorales, más abajo una célula con núcleo doble. El citoplasma está bien conservado. Las células corresponden por su tamaño a células basales

Carcinoma

116. Tumor cells from a mature
squamous cell carcinoma
Well-preserved cytoplasm, the
nuclei are uniformly large, the
nucleoli prominent
117. Two tumor cells from a
moderately mature, squamous cell
carcinoma
One cell contains a double nucleus
118. Tumor cells from an adeno-
carcinoma of the endocervix
Some cells have a suggestion of a
columnar shape. The cytoplasm of
other cells has disintegrated.
Some nuclei are huge and multi-
lobed

Carcinom

116. Tumorzellen aus einem aus-
reifenden Plattenepithelcarcinom
Gut erhaltenes Cytoplasma, die
Kerne sind gleichmäßig groß,
große Nucleoli
117. Zwei Tumorzellen aus einem
mittelreifen Plattenepithelcarcinom
Eine Zelle enthält einen Doppel-
kern
118. Tumorzellen aus einem Adeno-
carcinoma colli
Bei einigen Zellen ist die
Cylinderzellform noch angedeutet,
bei anderen Zellen ist das
Cytoplasma aufgelöst; monströse,
mehrlappige Kerne

Carcinoma

116. Células tumorales de un carcinoma
epidermoídeo maduro
Citoplasma bien conservado, los
núcleos son regularmente grandes,
nucléolos grandes
117. Dos células tumorales de un
carcinoma epidermoide de
mediana madurez
Una célula contiene un doble
núcleo
118. Células tumorales de un adeno-
carcinoma cervical
En algunas células la fórma
cilíndrica es todavía reconocible,
en otras células está el citoplasma
disuelto; núcleos monstruosos,
multilobulados

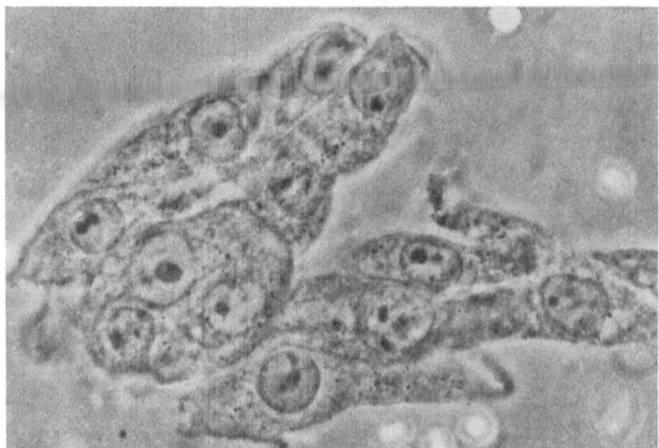

116

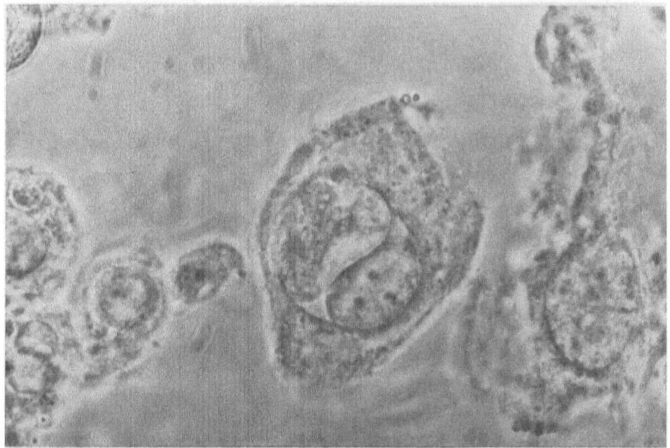

117

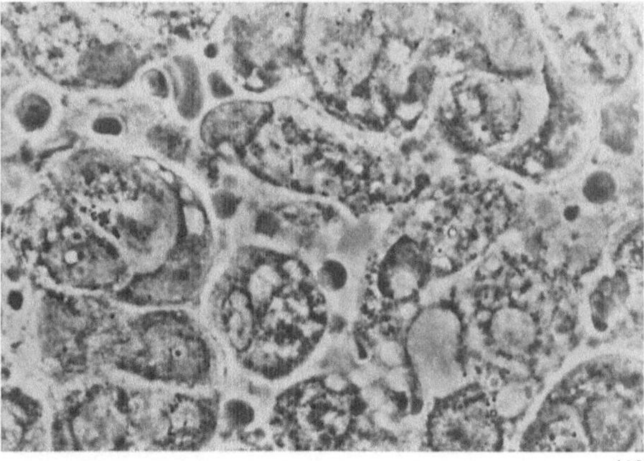

118

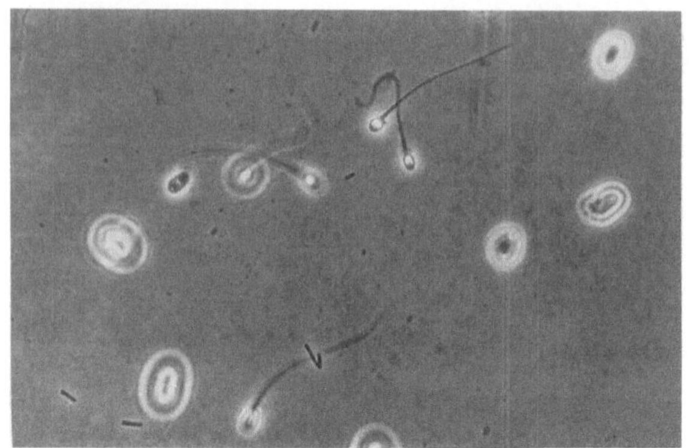

119

Sperm in Cervical Secretions

Secretions taken from the cervical os ten hours after intercourse

119. Phase-contrast

120. and 121. Interference contrast

Spermien im Cervicalsekret

Entnahme aus dem Muttermund 10 Std nach der Kohabitation

119. Phasenkontrast

120. und 121. Interferenzkontrast

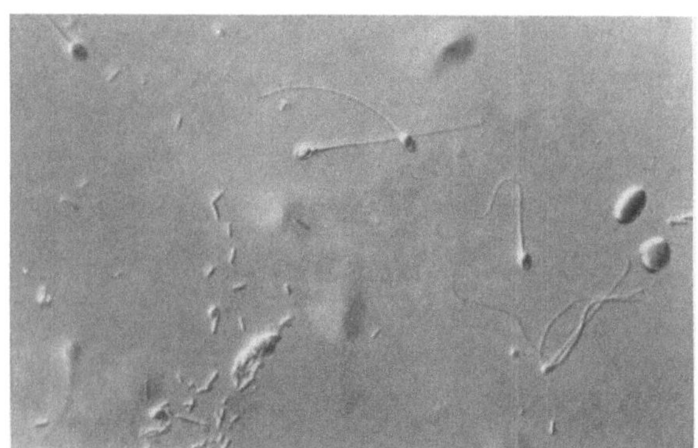

120

Espermios en la secreción cervical

Muestra tomada del orificio exocervical, 10 horas después de la cohabitación

119. Método de fases contrastadas

120. y 121. Método de interferencia

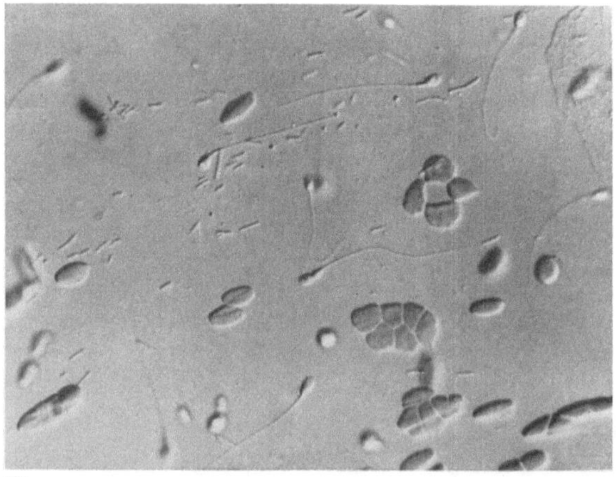

121

Cervical Mucus (the Fern-Leaf Test)

122. An intense fern-leaf test of the proliferative phase
tenth day of menstrual cycle

123. Slight crystallization of the secretory
phase
18th day of cycle

124. A suggestion of crystallization
during pregnancy
2nd Month

Cervicalschleim (Farnkrauttest)

122. Kräftige Farnbildung in der Proli-
ferationsphase
10. Cyclustag

123. Zarte Kristallisation in der Sekre-
tionsphase
18. Cyclustag

124. Angedeutete Kristallisation in der
Schwangerschaft
Mens II

Moco cervical (test de las hojas de helecho)

122. Intensa formación de hojas de
helecho en la fase proliferativa
10° dia del ciclo

123. Delicada cristalización en la fase
secretoria
18° dia del ciclo

124. Cristalización sólo insinuada en el
embarazo
2° mes de gestación

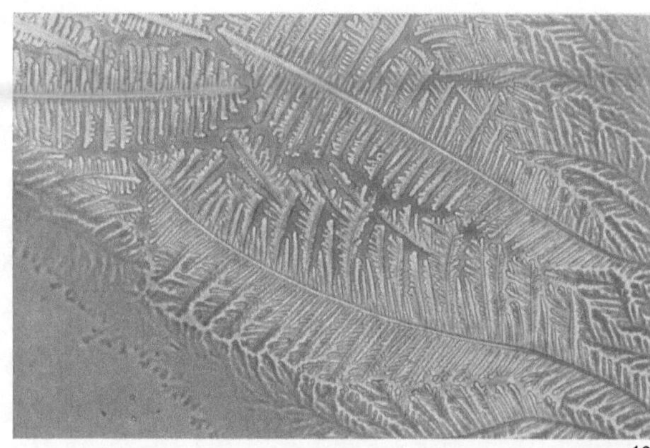

122

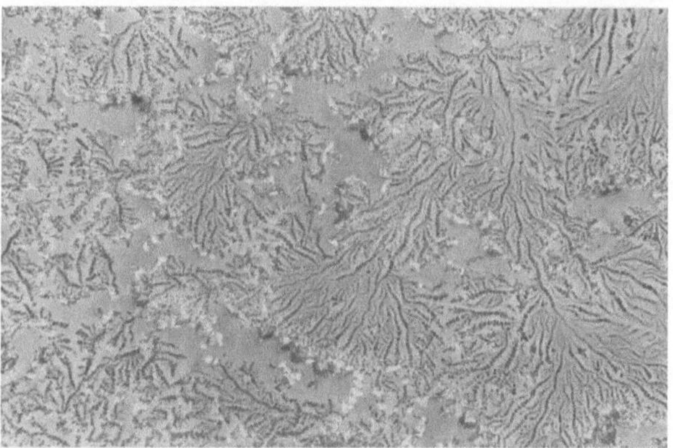

123

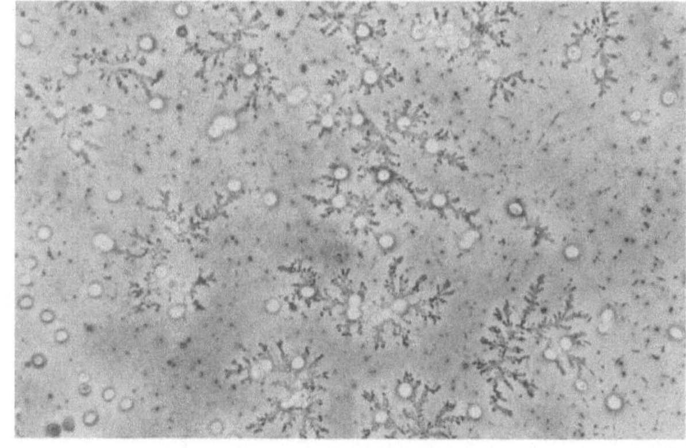

124

125

126

Artifacts

125. and 126. Dried smear

Comparison of phase-contrast and
interference-contrast

Artefakte

125. und 126. Austrocknung

Vergleich Phasenkontrast und
Interferenzkontrast

Artefactos de técnica

125. y 126. Resecamiento

Comparación entre contraste de
fases y interferencia

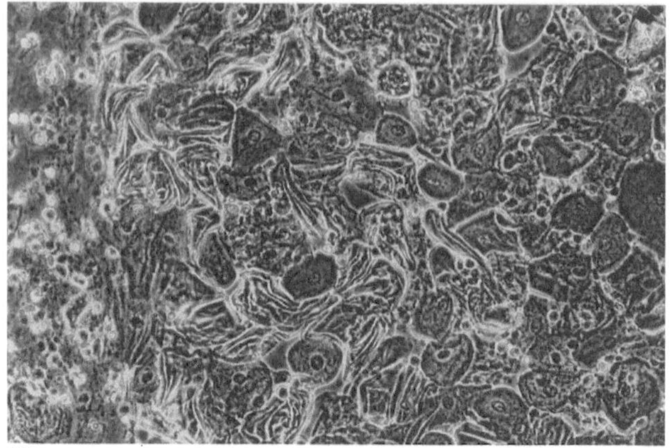

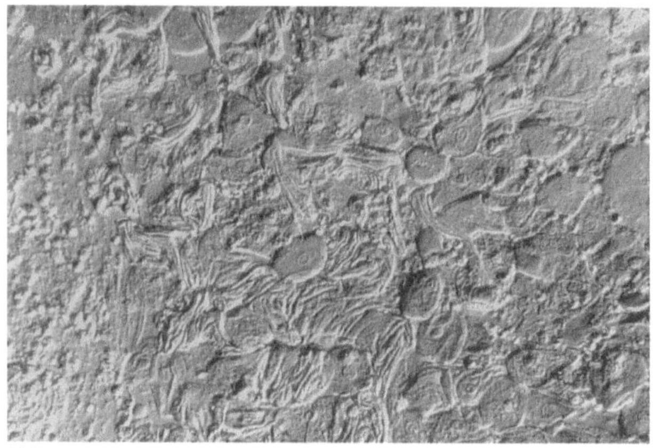

Artifacts

127. and 128. Effect of hypotonic
solution
Swelling of the poly-
morphonuclear leukocytes,
extrusion of cytoplasm.
Phase-contrast and
interference-contrast

Artefakte

127. und 128. Verweilen in
hypotoner Lösung.
Quellung der Leukocyten,
Plasmaaustritt.
Phasenkontrast und
Interferenzkontrast

Artefactos de técnica

127. y 128. Permanencia del
preparado en solución
hipotónica.
Imbibición de los leucocitos.
Dispersión del plasma.
Contraste de fases
y interferencia

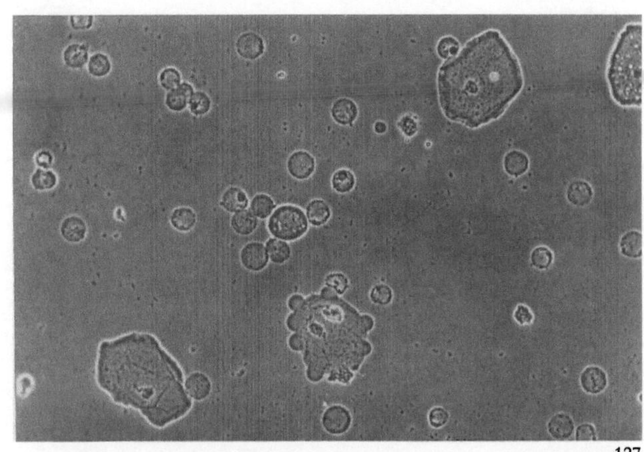

127

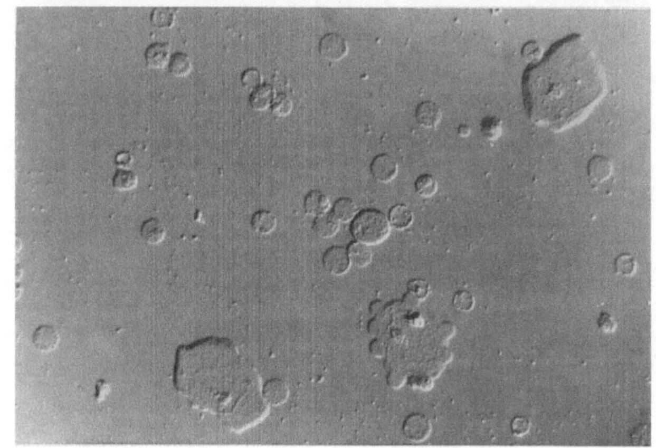

128

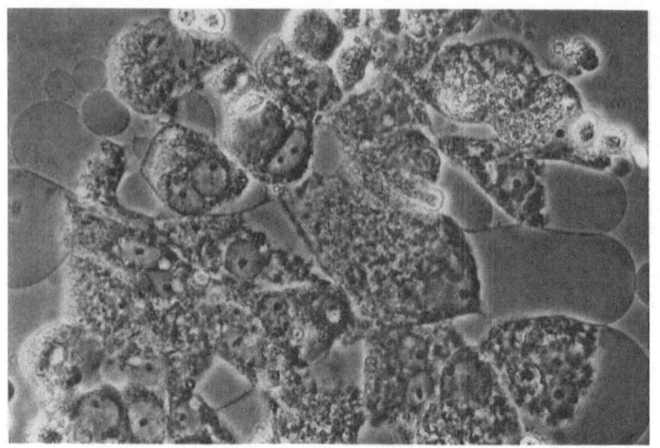

129

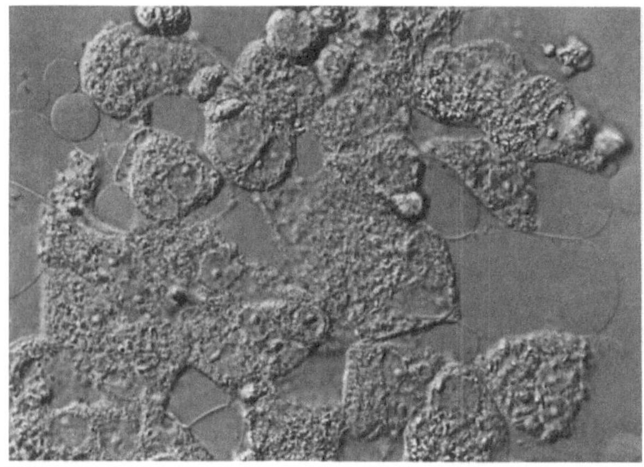

130

Artifacts

129. and 130. Endocervical
epithelial cells. Loss of water
and shrinkage

Artefakte

129. und 130. Endocervicalzellen
Wasserverlust und
Schrumpfung

Artefactos de técnica

129. y 130. Células endocervi-
cales. Arrugamiento celular
y pérdida de agua

Artifacts

131. and 132. Drying and
   agglutination of poly-
   morphonuclear leukocytes

Artefakte

131. und 132. Austrocknung und
   Anlagerung von Leukocyten

Artefactos de técnica

131. y 132. Resecamiento y
   aglutinación de leucocitos

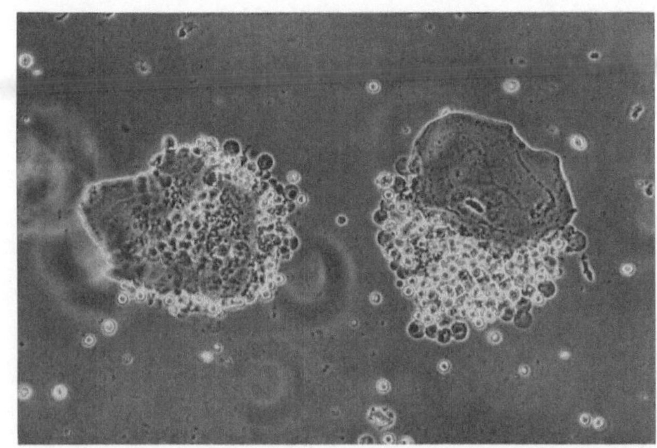

131

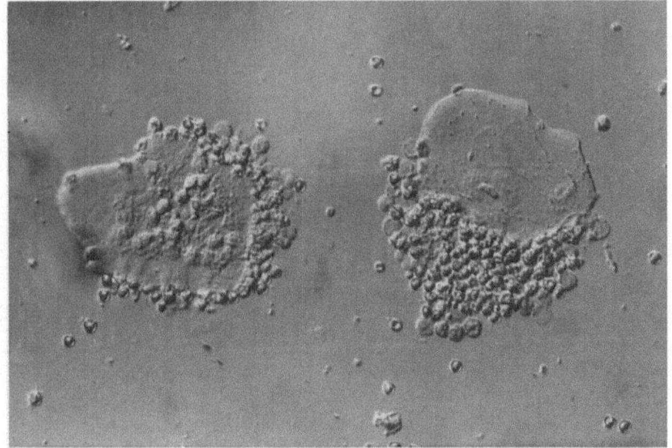

132

<table>
<tr><td>

Artifacts

133. Drying with loss of fluid
Swelling of the poly-
morphonuclear leukocytes

</td><td>

Artefakte

133. Austrocknung mit Flüssig-
keitsaustritt, Quellung der
Leukocyten

</td><td>

Artefactos de técnica

133. Resecamiento. Dispersión del
líquido plasmático.
Imbibición de leucocitos

</td></tr>
<tr><td>

134. Drying out

</td><td>

134. Austrocknung

</td><td>

134. Resecamiento

</td></tr>
<tr><td>

135. Swelling and formation of
vacuoles

136. Nuclear swelling and
disintegration of the
cytoplasm

</td><td>

135. Quellung und Vacuolen-
bildung

136. Kernquellung und Auf-
lösung des Cytoplasmas

</td><td>

135. Imbibición y formación de
vacuolas

136. Imbibición nuclear y
disolución del citoplasma

</td></tr>
</table>

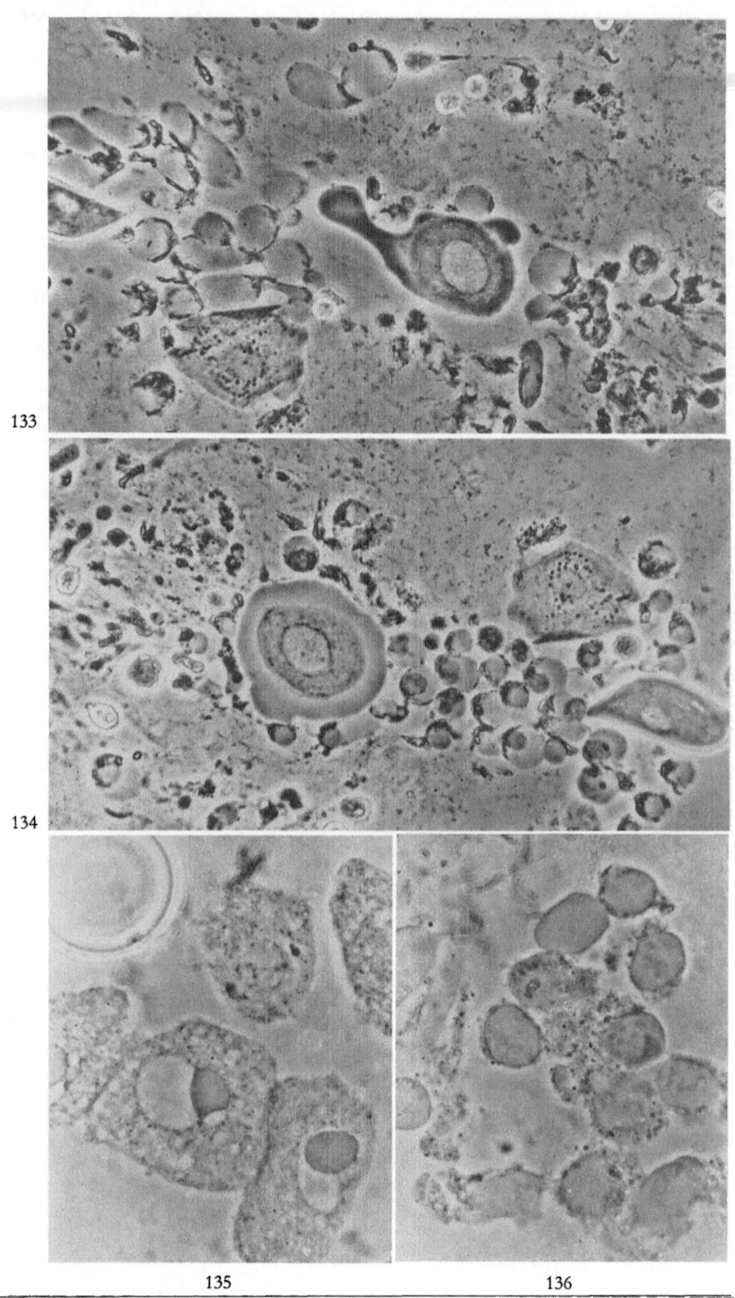
133
134
135
136

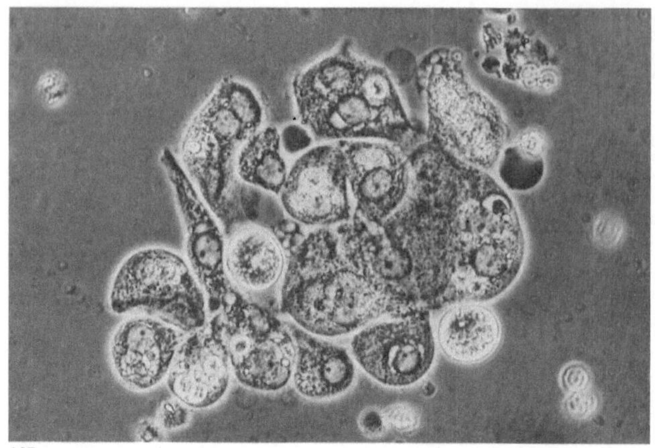

137

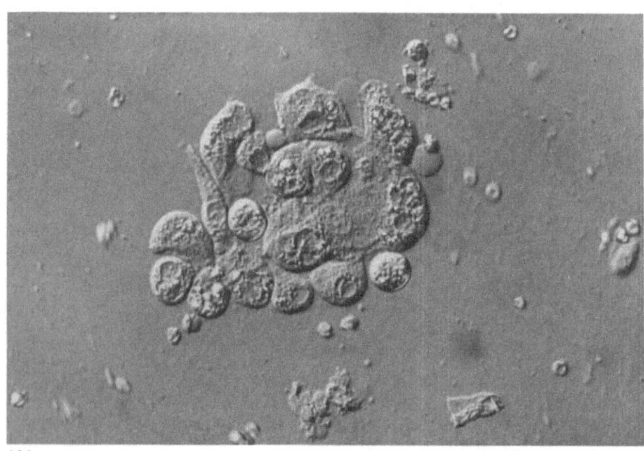

138

Artifacts

137. and 138. Dysplastic cells
with loss of turgor

Artefakte

137. und 138. Dysplastische
Zellen mit Turgorverlust

Artefactos de técnica

137. y 138. Células displásticas
con pérdida del turgor

Artifacts

139. and 140. Carcinoma cells
   with loss of turgor by
   efflux of cytoplasm

Artefakte

139. und 140. Carcinomzellen
   mit Turgorverlust und
   Plasmaaustritt

Artefactos de técnica

139. y 140. Células carcino-
   matosas con pérdida del
   turgor y salida de líquido
   plasmático

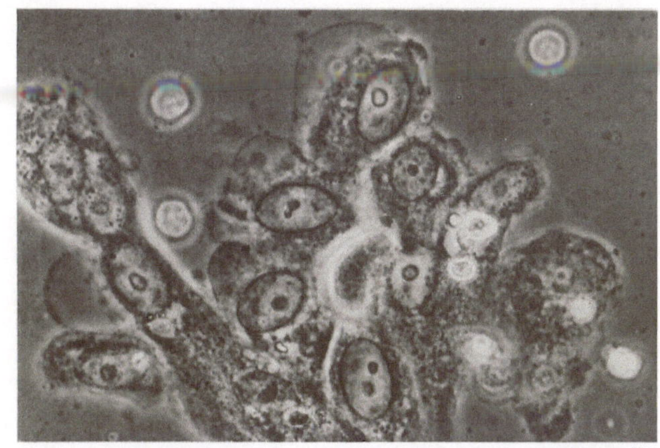

139

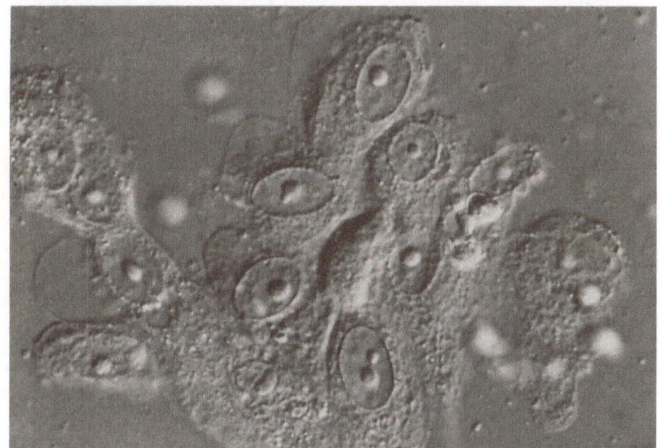

140